Praise for **Model Business Letters, Emails and other Business Documents**

Model Business Letters is one of the most pop ... ryone who borrows it sends back a glowing report. It's an exce ... ess – whether self-employed or working as a manager, secretary ...

Pamela Aitcheson, MIQPS, IQPS Training Li ... ate Secretaries, UK, **www.iqps.org**)

People who learn to write well are increasing their value to the company, helping to create and enhance the company's corporate image and playing a major part in its success. With so many executives and managers now composing their own correspondence, *Model Business Letters* contains essential guidance that will help people all over the world to upgrade their writing skills and thus help them to achieve these aims.

Joy Chan, Singapore

I have been searching for a simple yet effective book that deals with writing business letters and I have finally found one. *Model Business Letters* has been an invaluable help to me and I have, in fact, recommended it to Japanese buinessmen who wish to improve their written correspondence. The layout is simple and easy to follow and the models are in clear, plain English. The book is even being used as a text for non-native speakers of English, and both teachers and student speak very highly of it. I certainly advise anyone who wants to write professional yet uncluttered business documents to buy this book.

Michelle Sumura, Managing Director, 'Let's Go Australia', Perth, Western Australia

I enjoyed everything about Shirley's Powerful Business Writing Skills workshop. It is very lively, and Shirley was very willing to share her knowledge and always had time for questions. She also does not make us feel stupid for asking questions. With this 7th edition of her popular book, Shirley once again has another runaway international success that will become a valuable reference for people all over the world.

Lim Huey Lih, Head of Corporate Affairs, MSD, Malaysia

Congratulations on the 7th edition of your popular book, Shirley. I'm sure it will once again fly off bookshelves everywhere. It's right on target, and set to become the industry's benchmark for successful business writing.

Alan Hill, Managing Director, Ward Hi-Tech Ltd, Sheffield, UK

Booksellers tell me that while their shelves are full of books on business writing, it is *Model Business Letters* that sells and sells, so they never hesitate to stock it well. Not many other business writing books can claim to have sold almost half a million copies too. I once had to write an important and urgent press release for the launch of a new book. Without fail, I found just what I needed in *Model Business Letters*, and it did the job superbly. Well done on your fantastic 7th edition, Shirley.

Leslie Lim, Product & Sales Manager, Pansing Distribution, Singapore

Shirley, your workshops on business writing skills are fabulous. You are a natural presenter, making difficult things simple and bringing everyone into your world of writing. I have never had so much fun learning how to write better business letters! With *Model Business Letters, Emails and Other Business Documents* 7th edition, you have done the same thing. It's an easy-to-read, comprehensive and clear guide that will be indispensable for businessmen and women all over the world.

Ricky Lien, Trainer, Mindset Media, Sydney, Australia

In her workshops, Shirley has such a friendly and interactive approach that makes everyone feel comfortable. She has done the same in the 7th edition of her popular book, writing in a simple, clear way and giving practical examples. Well done on a great new edition, Shirley!

Carmelia Ng, Organisation Development Manager, Fuji Xerox Singapore

With our communication today becoming more and more faceless, our writing is more important than ever. Shirley's emphasis on building relationships through effective writing is really essential. In this 7th edition of her popular book, Shirley has given us a practical reference guide providing all the guidance we need as well as samples we can adapt. The new sections on email, social media and online writing are really useful too. Congratulations on another great success, Shirley.

Stephen Choo, PhD, Director and Regional Head of Insight, ASEAN, Hay Group

Shirley teaches in a practical and fun way how to ditch the dinosaur clichés that have no place in modern business writing. There is never a dull moment in her workshops. She has done the same with the 7th edition of her best-selling book *Model Business Letters*. It's a must-have for everyone who wants to write clear and persuasive business documents.

Alex Tan, HR Manager, Singapore

Model Business Letters, Emails and Other Business Documents

7th edition

Shirley Taylor

ORIGINALLY WRITTEN IN 1971 BY

L Gartside

Former Chief Examiner in Commercial Subjects, College of Preceptors

PEARSON

Harlow, England • London • New York • Boston • San Francisco • Toronto • Sydney • Auckland • Singapore • Hong Kong
Tokyo • Seoul • Taipei • New Delhi • Cape Town • São Paulo • Mexico City • Madrid • Amsterdam • Munich • Paris • Milan

PEARSON EDUCATION LIMITED

Edinburgh Gate
Harlow CM20 2JE
Tel: +44 (0)1279 623623
Fax: +44 (0)1279 431059
Website: www.pearson.com/uk

First published in Great Britain in 1992
Seventh edition published 2012

© Pearson Education Limited 2012

The right of Shirley Taylor to be identified as author of this work has been asserted
by her in accordance with the Copyright, Designs and Patents Act 1988.

Pearson Education is not responsible for the content of third-party internet sites.

ISBN: 978-0-273-75193-9

British Library Cataloguing-in-Publication Data
A catalogue record for this book is available from the British Library

Library of Congress Cataloging-in-Publication Data
Taylor, Shirley, Cert. Ed.
    Model business letters, emails and other business documents / Shirley Taylor. -- 7th ed.
        p. cm.
    Includes index.
    ISBN 978-0-273-75193-9 (limp)
    1. Commercial correspondence.   2. Business writing.   3. Letter writing.   I. Title.
    HF5726.T373 2012
    651.7′5--dc23
                                                                2011050142

10 9 8 7 6 5 4 3 2 1
16 15 14 13 12

Text designed by Design Deluxe
Typeset in 10/13.5pt Melior by 35
Printed and bound in Malaysia (CTP-VVP)

Contents

Go to **www.pearson-books.com/modelbusinessletters** where you'll find a selection of sample documents ready to download as templates for your own use.

About the author

Shirley Taylor has established herself as a leading authority in modern business writing and communication skills. Her books are popular all over the world, and have been translated into many languages.

Born in the UK, Shirley has also lived and worked in Singapore, Bahrain and Canada. She now lives in Singapore where she runs her own company. ST Training Solutions Pte Ltd has quickly become well regarded for providing high-quality public training programmes with professional trainers.

With 30 years' experience in teaching and training, Shirley knows what makes a great learning experience. She takes a personal interest in working closely with experienced trainers to ensure every workshop is practical and participative. With many repeat clients, consistency is important to Shirley, so she makes sure every workshop is full of valuable tools, helpful guidelines and powerful action steps.

Shirley specialises in training modern business writing skills, email etiquette, confident communication skills and success skills for secretaries. She puts a lot of passion and energy into her workshops to make sure they are entertaining, practical and informative, as well as a lot of fun. Shirley is also a popular, engaging keynote speaker at international conferences.

Shirley offers special reports and 10-week courses of Success Boosters in many communication-related topics. You can receive the special reports by signing up on her home page, **www.shirleytaylortraining.com**, or just click on Success Boosters to receive these helpful reminders.

Shirley's contact details are:
Website: **www.shirleytaylortraining.com**
Email: **shirley@shirleytaylortraining.com**

Publisher acknowledgements

We are grateful to the following for permission to reproduce copyright material:

Logos

Logos on page 58, page 62 from ST Training Solutions Pte Ltd, Singapore.

Text

The report on pages 451–64 from *The A-Z of Alternative Words*, Plain English Campaign.

The report on pages 465–87 from *How to Write Reports in Plain English*, Plain English Campaign (2001).

Extract on pages 275–7 from www.deskdemon.co.uk, with permission from Desk Demon.

Extract on pages 317–20 from *How to Write Proposals and Reports that get Results*, Prentice Hall (Jay, Ros 2003) 151–5 with permission from Pearson.

In some instances we have been unable to trace the owners of copyright material, and we would appreciate any information that would enable us to do so.

Preface to the first edition

By Leonard Gartside

Few business transactions are carried through successfully without correspondence at some point. Enquiries must be answered, quotations given, orders placed, complaints dealt with, transport and insurance arranged and accounts settled. Letters must be written to customers, salesmen, agents, suppliers, bankers, shipowners and many others; they cover every conceivable phase of business activity. They are the firm's silent salesmen and, often enough, represent its only contact with the outside world. Hence the need to create a good impression, not only of the writer's firm but also of the writer himself as an efficient person eager to be of service.

In the pages that follow are to be found over five hundred specimen letters dealing with a comprehensive range of transactions of the kind handled in business every day. They are represented, not as models to be copied, for no two business situations are ever quite alike, but rather as examples written in the modern English style to illustrate the accepted principles of good business writing.

Every business letter is written to a purpose; each has its own special aim, and one of the features of this book is its use of explanation to show how the various letters set out to achieve their aims. Basic legal principles relevant to different types of transaction are also touched upon, but only where there is a need to clarify legal relationships. Where the book is used in class, the letters provide material for teachers who may wish to enlarge on these matters and the exercises the means for students to apply in practice what they have been taught.

The many letters included are written in the straightforward and meaningful style of the modern age and should be of special help to the overseas user, and particularly to students in schools and colleges where commercial correspondence is taught either as a general business accomplishment or as a preparation for the various examinations.

November 1971
L.G.

Preface to the seventh edition

By Shirley Taylor

Take a look around your workplace. Do the leaders in your organisation write effectively and powerfully? Do good writers tend to get promoted? Do people tend to listen to good writers? Are good writers able to persuade or convince effectively? Absolutely they do!

More and more of our work today is undertaken through writing rather than in person or on the phone. As we are writing so much more these days, we depend on our writing skills to influence, persuade, encourage, collaborate, and to lead. However, how often do you notice people talking about the importance of good writing in your day-to-day work? They don't, do they? Most people don't really notice the quality of the writing they read – they simply react positively, negatively, or not at all. If you have ever wondered if there's a better way to write your messages so they get better results, there is!

Here are three of the new rules for written communication:

1 If you can say it, you can write it

We connect with the world today largely through email, websites, blogs, texting and social media. With all these channels we have only bare facts, without tone of voice, facial expressions, body language or pauses. As we regularly use these means instead of talking, it makes sense to use writing that's as close as possible to spoken language. When you do this, you gain yourself a great advantage – you put your personality and individuality into your message. This will help you to stand out more and make a greater connection with your reader.

2 Write for today, not yesterday

Yesterday's writing is passive and wordy, and it sounds really dull. It puts a distance between you and the reader. The way it is written also slows down understanding.

Yesterday:

Please be advised that a meeting of the Annual Convention Committee will be held on 24 February (Thursday) at 9.30 am. Approximately 2 hours will be required for the meeting and you are required to attend to report on progress made since the last meeting. Kindly advise me of your availability at your soonest.

Today's writing sounds more conversational. It's crisp, clear, transparent, and the personal context makes it more positive and interesting.

Today:

I'd like to hold another meeting of the Annual Convention Committee on Tuesday 24 February from 9.30 to 11.30 am. I hope you can attend to report on the great ideas you brought up at the last meeting. Please confirm if you can join us.

3 Aim to build relationships

In writing, as readers can't see or hear you, people will judge you based on what you write and how you write it. In today's fast-paced, communication-crazy world, it's essential to come across as a human being. If you insist on using old-fashioned or redundant jargon (*Please be reminded*, *Kindly be advised*, *Please find attached*, *herewith*, *above-mentioned*, *reference and perusal*, etc) you will obscure the real meaning and will not be adding any personality of your own. Make your writing positive, stimulating and interesting, add some feeling and a personal touch. This will help people get to know the real person behind the message.

Poorly written messages reflect badly on you and your organisation. Poor writing will not clarify your organisation's products, services, values, policies and beliefs; it may even portray them negatively. As a result, your reputation may be ruined and business efficiency lost. You will also lose out on opportunities to connect and to build relationships with clients, colleagues and collaborators.

Good writing is receiving increasing recognition as an essential business skill, and it will give you a huge advantage in today's business world. Good writing can help you work more efficiently, build credibility, improve relationships, influence other people, win more clients and achieve your goals.

If you want to give yourself an edge in this very competitive world, you must get to grips with effective writing now. And you have certainly found the right resource with *Model Business Letters, Emails and Other Business Documents*, seventh edition.

These three main rules for written communication are reinforced constantly throughout this book. Study its pages well and you will learn how to enhance your professional reputation and build great relationships.

Shirley Taylor
May 2012

Introduction

When you are talking to someone face-to-face, you have lots of visual clues to help you – your tone of voice, gestures, movements and eye contact. It may not be fair, but in the everyday world you are judged and influenced by all these criteria, and more – even your occupation, status, height, dress and the way you look. And like it or not, it's through the way you speak and look that you earn trust and confidence.

So, with none of these visual cues present in written communication, how do you earn trust and confidence when you write letters, reports, email messages or other business documents?

Earn trust and confidence by improving your writing style

In written communication you have to find other ways to evaluate the person who is 'speaking' (writing). People do that by looking at 'style'. Style means paying attention to proper spelling, punctuation and sentence construction.

Style also means:

- being creative in what you write and how you write
- making your communication look visually attractive by leaving a line space between each paragraph
- using numbered points or bullets appropriately
- using headings of a consistent style
- considering the appropriate tone in your writing
- structuring your messages logically.

I recently did a follow-up workshop for a client who had run my business writing workshop two months earlier. I asked the participants what had changed since our workshop. They told me proudly:

■ We get straight to the point, using everyday language instead of beating about the bush with old-fashioned, useless phrases.

■ Our messages are structured more logically so the reader can clearly see the action needed.

■ We try to avoid the passive phrases that we used to use, like *Please be informed, Kindly be advised, Please find attached*, etc.

■ We seem more approachable because our language is less formal and much more friendly, as though we are having a conversation.

Bingo! This is how to earn trust and confidence!

When conducting business today, using language effectively will go a long way towards achieving positive results. Good communicators go to considerable trouble to become competent in the English language, and it's often achieved gradually through a life-long learning process. Time, patience and hard work will bring enormous rewards and satisfaction, not to mention great results.

The language used in business today should be simple, courteous, relaxed and straightforward. Some of the key reasons why you need to work on your language in all your written communications are emphasised throughout this book. They are:

1 **To establish relationships**. People get an impression of you from the first email they receive, so it's important to make a connection by using appropriate words and phrases. For example, 'We spoke' or 'As spoken' will not have the same effect on your reader as 'Thanks for your call' or 'It was great to speak to you'.

2 **To communicate your ideas precisely**. Using unsuitable or incorrect expressions, or a long-winded writing style, will not give the reader the right meaning or the right impression. It will only lead to misunderstandings and lengthy correspondence to clarify.

3 **To convey a good impression**. Clear, concise, accurate language will give an impression of efficiency, and will fill the reader with confidence. Careless or inaccurate expressions will do the opposite. Readers may question if such carelessness will extend to other business dealings too.

Model Business Letters, Emails and Other Business Documents, seventh edition, will help you create a great writing style that will build trust and confidence in all your relationships. With 100 great tips for better business writing throughout the book, this is your one-stop shop, your desktop companion, the only guide you will need to help you to write any kind of business correspondence.

Who is this book for?

Many people will find this book useful:

- **Executives and managers who regularly compose their own correspondence**. Many managers now compose their own correspondence on their desktop, laptop or notebook. These ready-to-use documents can be copied or adapted to meet your precise needs. They will help you to say what you want to say and achieve the desired results. You will be able to save time and do your job better, more effectively and easily, without scratching your head for ages thinking about where to start or what to write.

- **Overseas users**. Past editions of this book have been sold extensively from India to Indonesia, from Malaysia to the Maldives, from Singapore to Sri Lanka, from Hong Kong to Harrogate, from Shanghai to Sheffield, from Toronto to Thailand. Overseas users will appreciate the value of this comprehensive resource book. It will be especially useful in dealing with international business transactions using modern business language.

- **Students and lecturers**. Students following a business, professional, secretarial or administrative examination course often need to compose business letters and other business documents. You will find the guidelines, theory, specimen documents, four-point plans, definitions and checklists particularly useful in learning how to compose your own effective business correspondence.

How is the book organised?

I have organised the book so that you can hop around to the topics that interest you. Just dip in whenever you face a blank page or screen, and hopefully you will find some inspiration to get you started and some useful tips to help you finish the task.

This seventh edition has been completely restructured to make it much easier to find documents.

Part 1 Communication: An Overview. This is a must for everyone. Chapter 1 discusses the importance of building relationships in business today. It's no longer enough to get the job done. You have to develop great relationships first. In Chapter 2 you'll find a new section looking at sentence construction. Many people asked me for some basic advice on this, and I agree that it's necessary. After all, how can I expect you to write an email unless I first of all give you advice on how to use proper punctuation and sentence construction?

Part 2 Business Writing Basics. In this part you will find all the fundamentals of modern writing. Chapter 3 is a completely rewritten chapter on the importance of using 21st century business language. In Chapter 4 regular readers will be pleased to see my four-point plan here again – it's a favourite formula that helps people all over the world to structure documents logically. Finally in Chapter 5 we look at attractive presentation and display of business documents.

Part 3 Electronic Communication. This is a brand new part going into much more detail on electronic communication. No book on writing would be complete without a detailed look at email etiquette, which you'll find in Chapter 6, including lazy email habits and top tips for making the most of email. In Chapter 7 we look at writing for websites, blogs and social networking and in Chapter 8 customer care online.

Part 4 Routine Business Transactions. This part contains just that – enquiries, quotations, orders, invoices, all the correspondence making up a standard business transaction.

Part 5 General Business Correspondence. Here we have another completely new part, with examples of internal correspondence, secretarial correspondence, meetings documentation, personnel correspondence, reports and proposals.

Part 6 Creative and Persuasive Documents. In this final part, we focus on documents that require creativity and persuasion, documents that need to influence and convince in some way. A wide variety of documents are featured, including the specific writing skills unique to each one, such as complaints, goodwill messages, notices and advertisements, circulars, sales letters, press releases and business plans.

Appendix. You'll find the Appendix very useful. It contains spoken and written forms of address, plus a new section that I know you will love, courtesy of the Plain English Campaign: The A–Z of alternative words.

Special features

- As each new document is introduced, the format is illustrated in a specimen document, with notes highlighting every aspect of document presentation.
- Each section contains full explanations, discussion and theory regarding the various documents.
- Many specimen letters are boxed and include marginal notes that discuss important features of the text.
- Four-point plans encourage you to plan and structure your own documents effectively.

Look out for these special icons:

 TIP Throughout the book you will find over 100 top tips for effective communication.

 CAUTION You'll find some words of warning here, such as things to avoid or precautions to take.

 CHECKLIST At the end of most chapters there is a checklist to remind you of the key points to remember.

Final word

A workshop participant once said to me, 'Rarely does a day go by that I don't pick up your book, browse through it, and find something that helps me with my writing.' That's what this book is all about.

I hope this reference book, with its emphasis on high-quality presentation, structure, language and tone, helps you to convey your messages appropriately and effectively. Remember that in so doing you will not only be helping to create and enhance the corporate image of your organisation, you will also be increasing your value to the company and playing a major part in its success.

By picking up this book you have already shown a desire to learn more about modern business writing. The solid advice and practical guidelines, combined with hundreds of sample documents, will show you how to develop effective written communication skills. The rest is up to you.

Good luck!

Shirley Taylor

PLEASE NOTE:

1 For reasons of consistency and simplicity and to avoid confusion, we have used the UK -ise spelling convention throughout this book. Readers should also be familiar with the -ize convention used in many countries worldwide.

2 The author and publishers feel that the advice in this book is sound. However, readers must seek legal advice if they are in any doubt.

part

Communication:
An Overview

'Coming together is a beginning; keeping together is progress; working together is success.'

Henry Ford

When I first started my working life, snail mail was the only mail. Telex was the main method of sending a message over any distance. The introduction of fax made a big difference to the speed with which we communicated. Then along came email and instant messaging, not to mention mobile phones and text messaging.

Our working environment is in a constant state of change, and no doubt that will continue at an even faster rate. If we are to meet the numerous challenges this changing workplace will bring, we will need to work even harder on our communication skills. Here are some of the reasons why:

■ **Advancing technology**. The Internet, email, fax messages, voice mail, instant messaging, teleconferencing, videoconferencing, wireless devices – all this technology has transformed the way we communicate. People can work together almost effortlessly whether they are in Sydney or Seattle, Shanghai or Sheffield, Bali or Brazil; and they may be in a hotel, at home, at an airport or on an aeroplane. With every phone call or email, your communication skills are revealed for everyone to see.

■ **Global communication**. More and more businesses are now working on a global scale across national boundaries. Many people now work for multi-national companies, and today's workforce all over the world now includes increasing numbers of people from different ethnic backgrounds. If you are to communicate effectively in this environment, you must understand other people's backgrounds, beliefs and characters.

■ **The information age**. With an increase in the amount of information in today's business world, you must be able to make quick, effective decisions based on the information you receive. You must also know how to find, assess, process and communicate information efficiently and effectively. With so much information available today, it is a constant challenge to get your recipient's attention so that they will read and act appropriately on your message.

■ **Team-based business environments**. In today's fast-paced business world, the traditional management hierarchy has changed, and team working is now in vogue. In such a team-based environment it is important to study and understand how groups work together. You must learn to listen and watch other people carefully so that you interpret all the non-verbal cues you receive.

 TIP Learning more about effective communication will help us all adapt to changing environments, and will increase your chances of success in this highly competitive world.

The importance of building relationships – both orally and in writing

No one has a job that requires them to work alone. We must all interact with various levels of staff – with bosses, managers, supervisors, colleagues and co-workers, and with junior staff. We liaise constantly with internal and external customers. The key to your success in your job and in your career will not be gaining more paper qualifications. Your success will depend on your ability to develop relationships.

Communication is our lifeblood, and it's the lifeblood of any organisation. Excellent communication skills are probably the most important career and personal skills you can possess. But people aren't just 'born' writers or speakers. The more you write and the more you speak, the more your skills will improve. Of course, you must also take an interest in learning how to improve, and take action to do so.

Improving your communication skills will put you ahead of other people when it comes to getting a job, gaining promotion, even just getting your job done. More importantly, developing your communication skills will help you to build successful relationships both personally and professionally.

 TIP The greatest challenge you face will probably not be the technical side of your job (your expertise) but rather interacting with other people.

Dangers of poor communication

Let's take a look at some of the things that could go wrong when we don't work on our communication skills:

When communication breaks down, it has not only an immediate effect on you and the other person, but also on your whole team. A breakdown in communication means a link in the chain is broken. There can be far-reaching effects throughout the department and even the organisation.

 CAUTION Make it an on-going process to work on your communication skills. It's better to start now rather than wait until something goes wrong!

Successful people develop great relationships and use them to build a foundation for success

Every day we interact with dozens of people, whether it's on the phone, face to face, along the corridor, in the lift, in group discussions, meetings, or of course in writing. All outcomes are determined as a result of interactions with others. If people avoid you, it will be impossible to get work done. Your professional effectiveness at work will largely, therefore, depend on how much people are willing to interact with you.

Successful people consider relationships not as a means to an end, ie to get a job done. They understand that strong relationships create loyalty, build trust and instil confidence, and are ultimately the key to success.

 TIP In any job or business, relationship building must be your most important objective. Quite simply, the quality of the relationship will determine the quality of the product or service.

Seven key factors to building great relationships

If you want to build better relationships, at home and at work, orally and in writing, here are seven key factors you need to work on.

1 Be courteous

I see and hear of people who walk into the office each morning with their eyes down, headphones in their ears. They don't even glance at others, let alone offer a simple, 'Good morning!' Others can be downright abrupt and disrespectful to colleagues and subordinates. It's no excuse to justify this by complaining of pressures of work. We are all busy! Everyone has a right to work in a cordial environment, and work flows more smoothly when the atmosphere and the people in it are pleasant. Courtesy is the oil that keeps the engine of any relationship running smoothly.

2 Find common interests

How many people in your office do you really know? How many times do you enter the lift with the same person yet never even acknowledge them? Do you walk past co-workers' desks and never nod your head or say hi? What a sad way to work. If you make an effort to get to know your colleagues and clients, you can then build on commonalities. For example, comment on a photo or an object on a colleague's desk. You may find you have a story to share, or you may learn something new that you can discuss. Making an effort to gain eye contact, spark up a conversation, smile, even just nod and say 'hello' is also a much more enjoyable and rewarding way to spend your day.

3 Build credibility

Very often at work you will have to convince people of your point of view. You need credibility for this. You will gain a certain amount of credibility from your experience. However, if you are to build strong connections – connections that you can count on when you have new ideas and goals – you need to gain respect, create trust and build rapport. You can do this through being credible. Do you turn up for meetings and share your knowledge? Do you keep others informed? Do you do what you said you'd do in a timely manner? Are you honest? Credibility comes with transparency, engagement and honest hard work.

4 Make others feel important

I'm often surprised that some people have a lot of trouble acknowledging good work, or saying a simple 'thank you' for a job well done. Feeling unimportant or unappreciated is extremely demotivating. If you are a manager, make an effort to talk to your staff about something other than business from time to time. Ask them about their families, their upcoming holiday, their weekend. Listen to them. Show you are approachable. By doing this you will win their respect, and at the same time you'll learn more about your staff and will pick up useful information that could help you guide and motivate them.

One of the most fundamental rules of developing relationships is to respect other people's feelings. We all like to be recognised and appreciated. If you want to make friends and enhance your reputation as a great communicator, learn how to make others feel important.

5 Show humility

There's nothing worse than someone who brags and boasts about themselves, with their nose in the air and an air of arrogance. These people will have others running away from them rather than wanting to get closer. Humility involves maintaining our pride about who we are and about our achievements, but without arrogance. Humility means having a quiet confidence and being content to let others discover your talents without having to brag about them. Interestingly, very often the higher people rise and the more accomplishments they have, the higher their humility index.

 TIP If I want to improve my relationships, I must practise humility. It's a strength, not a weakness.

6 Listen actively

Take an interest in other people by listening to them. You may learn some useful information that you can use to create value in the future. Chat to a client about his family. Find out your boss's likes and dislikes. You never know when this information may be useful. For instance, when I was a secretary I heard the boss's wife commenting to someone on how she liked a certain perfume, so when it came to her birthday, and the boss didn't know what to get her, guess what I suggested? Flatter someone today by getting to know them better through active listening.

7 Be empathic

Empathy is all about getting to know people and understanding how they feel. The need to be understood is one of the highest human needs, but many people don't care or just don't make an effort to find out how we really feel. In my Communication workshops, I find people have a lot of difficulty with empathy or expressing real feelings. I think this is a shame. Just imagine the difference you can make if you really get to know people and understand how they feel. It could really set you apart from the rest and you'd start giving great value that many others don't give.

 TIP Think about one of your best relationships at work. Why does it work so well? Why is it successful? Is it because of one (or more) of these seven factors?

Most people would agree that their satisfaction at work is largely derived from the way they, their colleagues and their clients communicate. As with any other endeavour, the more you put into it, the more you'll get back. When you start practising these basic success tools for making great connections, you will see the massive rewards they can bring, both personally and professionally.

Communicating across cultures

Different cultures communicate differently. In today's offices we are all communicating much more not just with different countries, but with different nationalities within our own country. As such, it's a good idea to appreciate and try to understand each other's differences, recognising that not everyone thinks the same way that we do. You may indeed have to adapt your writing style according to the cultures with which you do business.

In all my workshops and books, I advocate a style of business writing that is simple, clear, concise, reducing verbosity at all costs. However, if you are to write fewer words, it's even more important that you get those words right.

Challenges of non-native English writers

No matter what your nationality, communication is always a challenge – that of conveying your meaning successfully to another person or persons. When we have to convey our meaning in writing, the challenge is even greater. You have to ask yourself if these are the right words to use in writing, considering you are not there to explain them to the reader.

For non-native English speakers, there are even more challenges. They also have an extra step in the process – they have to translate their words from their native language into English before they write them down. The danger here is that simple translations could become:

- overcomplicated or wordy
- focused on specific words rather than on the whole meaning
- lacking in any action needed.

As such, here's a systematic sequence for you to use in your approach:

1 Identify the thought in your own language.
2 Translate it into English.
3 Convert the thought in English into the correct written English words.
4 Consider if your reader is likely to interpret these words accurately.
5 Reflect on any changes that could be clarified before finalising.

If you are to avoid or minimise distortions, it's important to go through this sequence carefully and ask yourself such questions as:

- Will my readers know and understand the words I use?
- Are the words the right words?
- Have I stated the response I expect?
- Will my message achieve the desired results?

Keep it simple

Short, simple sentences are much easier to understand by someone who is hearing or reading messages in a language other than their mother tongue. Keep your messages as short as possible without letting the meaning suffer. Niceties are acceptable, to a point, but don't overdo it. Remember too that they are usually idiomatic and difficult to understand by someone not absolutely fluent in your language.

When crossing international boundaries in your email messages, it is better to err on the side of caution and use a more formal tone for your messages at first. It will then be easy to progress from formal to friendly as you get to know your recipient better. It would certainly weaken your position if you have to step backwards from friendly to formal.

How simple is simple?

A friend told me recently that she received a business email message written almost entirely in text language (for example, u r, wud, cud). When my friend replied she told the writer that he hadn't come across very positively because his message was too casual. The writer wrote back immediately in more appropriate language, and he thanked her very much for mentioning this. The writer had just chosen the wrong way of speaking (or writing) to my friend initially. The key is knowing when to use abbreviations and when not to.

 CAUTION When texting or emailing friends, or putting posts up on your Facebook for friends page, then perhaps abbreviated words are fine. But don't get into the habit of doing this so that you forget to spell out the words correctly when necessary.

 TIP You can read more about this in Chapter 7 on Writing Online: websites, blogs and social networking.

Remember your ABC

Good communication, both oral and written, results when you say exactly what you want to say using an appropriate tone. Always ask yourself if your message meets these essential requirements:

Accurate	Check facts carefully
	Include all relevant details
	Proofread thoroughly
Brief	Keep sentences short
	Use simple expressions
	Use non-technical language
	Use active voice
Clear	Use plain, simple English
	Write in an easy, natural style
	Avoid formality

Checklist

- Make a point of studying and continually developing your communication skills. It will reap many rewards.

- Never underestimate the power of building successful relationships.

- Be aware of the dangers of poor communication, and avoid communication breakdowns.

- Remember that a breakdown in communication between two people can affect the whole team.

- Increase your credibility by gaining respect, creating trust and building rapport.

- Understand the differences between cultures, and recognise that not everyone thinks the same way as you.

- Avoid misunderstanding orally and in writing by using simple, clear, concise language.

- Consider your audience when communicating, and use language they will understand.

- Keep your language simple, but not too simple that it would come across as unprofessional.

- Only use abbreviations and acronyms when writing to your friends, not your business contacts.

2
Sentence construction matters

D on't listen to anyone who tells you it's not important to use full sentences or get your grammar right – both orally or in writing. It is! Such errors can lead to misunderstanding, confusion and breakdown in communication. Incorrect punctuation or grammar can also lower your credibility as well as your organisation's image.

Many people make grammatical mistakes because they do not understand the rules properly, or simply through carelessness. Don't be scared of getting to know the various parts of speech and how they all work together. It may help you in your quest for better writing.

Grammatical terms

Noun A word that is the name of a person, place or thing.

Examples:

house, bird, office, desk

The *book* is on the *shelf*.

The *audience* participated in the *seminar*.

Collective noun A word used in the singular to express many individuals.

Examples:

furniture, committee, crowd, equipment, baggage, luggage

 TIP Always use a singular verb with collective nouns, eg The furniture is . . . , The equipment was . . .

Verb A word that expresses action or condition.

Examples:

I *wish* I could *go* to Australia.

The computer *developed* a fault.

Infinitive The verb form introduced by 'to'.

Examples:

to walk, to go, to finish, to play

Mrs Lim wants *to speak* to her daughter.

Adjective A word that describes a noun.

Examples:

thin, rare, red, long, pretty, amusing

The *boring* lecture ended at 11.30.

The *new* furniture will arrive today.

Adverb A word that gives more information about a verb. It also describes adjectives and other adverbs.

Examples:

happily, severely, frequently

The woman worked *conscientiously* at her job.

The man studied *hard* to pass his exam.

Conjunction A connecting word that links other words (or groups of words) together.

Examples:

and, or, but, as, when, than, after

The car had just started *when* smoke came out.

I like the designer suit *but* I can't afford it.

Interjection A word that expresses exclamation.

Examples:

Ouch! Wow! Gosh!

'Help!' she shouted.

Preposition A word that expresses the relationship between a noun and another part of the same sentence.

Examples:

at, by, on, for, to

Tom is sitting *on* the chair.

We must break for lunch *at* 12.30.

Pronoun A word that takes the place of a noun.

Examples:

I, you, we, their, that, either

She hopes to join a band.

Their new record is at number one in the charts.

Sentence A group of words that states a complete thought.

Examples:

Sylvia is married.

Wendy cried.

Subject This is the person, place or thing that you are speaking about.

Examples:

The *consultant* charged a fair price.

Sex and the City is a popular TV show.

Clause A group of words that includes a verb and forms part of a sentence.

An *independent clause* expresses a complete thought. It can stand alone.

Example:

Where's my shirt?

A *dependent clause* does not express a complete thought. It cannot stand alone as a sentence.

Example:

After the seminar has finished . . .

 TIP Read sentences out loud to hear pace and rhythm. This will also help you to get the punctuation in the right place.

Subject and verb agreement

The first rule in English language is to get your sentence structure right. A sentence is a group of words containing a complete expression, thought or idea. It must contain a subject and a verb.

The manager needs your report urgently. Sylvia and Diego were married today.
   | | |
   subject verb   subject   verb

In a longer sentence the subject and verb may not be next to each other. In these cases it's important to find the subject first and make sure you choose the verb that matches it.

The file containing the important documents was found on Jane's desk.
  | |
  subject   verb

In this example it would be easy to use the plural verb 'were' because the verb follows the word 'documents'; but the subject of the sentence is actually 'file' so we need to use the singular verb 'was'. Let's look at some more examples:

The use of computers in homes has increased a lot in recent years.
  | |
  subject   verb

Communication between people in different countries has speeded up with email.
   | |
   subject   verb

 TIP You will be less likely to make subject–verb errors if you keep your sentences short.

Punctuation

One of the most common challenges in business writing today is getting commas and full stops in the right place. I see so many commas where there should really be full stops. The thing is, when you put a comma (when it really should be a full stop), I'll read along the sentence, then after the comma I'll figure out it just doesn't make sense, so I have to reread from the beginning again to try to make sense of your sentence. This is all taking up too much of the reader's time. Study some simple rules for the comma and you may help yourself as well as your readers.

Here are the main rules for the comma:

1 Use a comma to separate words or phrases in a list

Would you like a gold, grey, white or black trim on your new car?

Writing well takes time, effort and a lot of hard work.

(NB: In a list like this, it's not necessary to put a comma before the final 'and', although in the US it is common to do so.)

2 Use a comma to separate adjectives qualifying the same noun

Please send us a large, self-addressed envelope.

I enjoy the warm, humid climate in Singapore.

3 Use a comma to separate two clauses that are joined by a co-ordinating conjunction (like *but*, *or*, *yet*, *so*, *for*, *and* or *nor*)

The expansion of our business is a long-term project, and we need an efficient management consultant to help us.

John has the necessary qualifications, but Dave has more experience.

I believe this candidate has the relevant qualifications, and he also has considerable past experience.

(NB: Whether or not you use a comma here will depend on your own preference, the length of the sentence and the amount of separation that you wish to show. In a short sentence where the ideas in the clauses are closely related, you might leave out the comma. In a longer sentence you might put in the comma so that the reader can better absorb what he or she has read so far.)

4 Use commas to create parentheses, where something is inserted that either expands on the main sentence or qualifies part of it

Mandy Lim, my secretary, will contact you soon to make an appointment.

The Managing Director, who is overseas at present, has asked me to reply to your letter.

I need Mark, and possibly Doreen as well, to help with this project.

Please switch off all electronic devices, including mobile phones, as the aircraft will be departing soon.

Unfortunately, this rule causes many problems. Let me show you some more examples:

> James, our Vice President has left the company.
>
> *In this sentence, the writer is telling James that their Vice President has left the company.*
>
> James, our Vice President, has left the company.
>
> *Here, the writer is telling someone else that James (who is the Vice President) has left the company.*

5 Use a comma to separate phrases and clauses to make your message easier to read

> We have five different models, each with its own special features.
>
> Although I agree with the points you mention, I would like clarification on various issues.

Using commas is largely a matter of taste and style, but one thing is for sure – they should not be overused. When I first revised this book in 1992, many of the letters were full of commas and very long sentences. Take a look at this example:

> Unfortunately, if we invest in new machinery, and the market falls again, as it has been predicted, we may, possibly, find ourselves with too much production capacity, and this may, therefore, result in even more serious problems.

While all the commas in this sentence are placed correctly, there are far too many of them and they make the sentence jerky. It's also way too long. In today's business writing we should keep sentences short and cut out non-essential commas. For example:

> We must give serious consideration to the issue of investing in new machinery. If the market falls again, as has been predicted, we could find ourselves with increased production capacity. This may then result in even more difficulties.

 TIP A comma represents a short pause. When reading any written messages, don't just see words on paper (or screen) – imagine you are reading the sentences out loud, as though you are speaking to your audience. This will help you to place the commas and full stops correctly.

Common errors with commas

A comma splice is when a comma is used between two main clauses where really a comma is not enough. Take a look at this example:

It was great to see you last week, thanks for your hospitality.

This is not correct. What we have here are two full sentences, two complete thoughts, so the comma in the middle should be a full stop:

It was great to see you last week. Thanks for your hospitality.

Another clue here is the word 'Thanks', which will always be the first word of a sentence, unless it follows a conjunction like 'and', 'so', 'but', etc, as we see here:

It was great to see you last week, and thanks for your hospitality.

Here are some more examples of this comma splice and how the sentences should be corrected:

Wrong: I ordered more disks, they will be delivered next week.

Right: I ordered more disks, and they will be delivered next week.

Or: I ordered more disks. They will be delivered next week.

Wrong: Please look into this immediately, I need your report by 28 November.

Right: Please look into this immediately. I need your report by 28 November.

Or: Please look into this immediately and let me have your report by 28 November.

Wrong: We have removed clause 9, this is because of current market conditions.

Right: We have removed clause 9 because of current market conditions.

Or: We have removed clause 9. This is because of current market conditions.

 CAUTION Putting a comma in the wrong place is one of the most common problems I see in today's writing. Read sentences out loud to hear the pace and rhythm. This will help you to get the punctuation in the right place.

Apostrophes

If you have difficulties with the apostrophe, you are not alone. Let me try to make this simple for you:

Contractions

In formal business letters we would write *I will be visiting Beijing next month.* When speaking we would say *I'll be visiting Beijing next month.* So it makes sense to use contractions like this in email writing. An apostrophe is used to indicate the missing letters (or sounds).

I'm (I am)	he's (he is)	you're (you are)
don't (do not)	haven't (have not)	wouldn't (would not)
it's (it is)	we'll (we will)	who's (who is)

Many people find it difficult to determine when to use *it's* or *its*. *It's* is a contraction that is always short for '*It is*' or '*It has*'. Take a look at these examples:

It's very tough working out where to put the apostrophe.

The cat's whiskers are on either side of *its* face.

It's very amusing when you see a little puppy chasing *its* tail.

The group made *its* decision.

 CAUTION There is no such thing as *its'*.

 TIP If you're unsure, try replacing *it's* with *it is*. If it sounds OK that way, then use the apostrophe.

Possession

When the owner is singular, we indicate possession by using an apostrophe followed by the letter 's'.

The applicant's letter

The director's office

The visitor's car park

The dog's bed

The girl's bicycle

When the owner is plural, we indicate possession by putting an apostrophe after the final 's'.

The applicants' letters

The directors' meeting

The visitors' car park

The dogs' beds

The girls' bicycles

When a word is plural without the letter 's' at the end, then we simply add the apostrophe and 's', as in these examples:

child's toy   children's toys   the geese's story

man's coat   men's coats   the people's pharmacy

 TIP When figuring out where to put the apostrophe when something is possessive, try putting a circle around the owner. Your apostrophe should not be inside the circle.

Some words of caution with the apostrophe

1 We don't need an apostrophe in the plural form of numbers and dates:

the 1920s   the roaring twenties

2 We don't need an apostrophe in possessive pronouns:

his   hers   its   ours   yours   theirs

3 When names end with the letter 's', either use is acceptable. I prefer to use the examples in the left column, because this is more in line with how the word is pronounced:

James's wife   James' wife

Douglas's car   Douglas' car

Iris's report   Iris' report

My boss's office   My boss' office

4 You don't need apostrophes when making plurals from capital letters:

Mary learned her ABCs today.

Checklist

- Remember that becoming a good communicator is a life-long learning process. It won't happen overnight.

- Enhance your credibility by paying attention to grammar and sentence construction.

- Get to know the various parts of speech and how they work together.

- Get your sentence structure right, and make sure your verb matches your subject.

- Read sentences out loud to hear pace and rhythm, and help you get the punctuation in the right place.

- Study the rules for the apostrophe and ensure they are placed correctly.

- Help yourself as well as your readers by keeping your sentences short.

- Convey an impression of efficiency by using clear, concise, accurate language.

- Make a great first impression from the very first email you send.

- Use relationship-building phrases like 'Thanks for your call' instead of relationship-breaking phrases like 'We spoke'.

Business Writing Basics

'Letters have to pass two tests before they can be classed as good: they must express the personality both of the writer and of the recipient.'

E.M. Forster

If you are like most people, you spend a lot of time every day tapping away at your computer, not just on emails, but also letters, fax messages, reports, proposals and much more. Have you ever wondered why you didn't get the reply you wanted after writing a message? Have you ever wondered how you could have expressed yourself better? Or how you could have achieved better results? The bottom line is this: there is a better way. In this part, I'll show you how.

Whoever you are and whatever you write, your writing can often mean the difference between success and failure. If your writing is full of standard template phrases that show no feeling and no personality, you are telling your reader you can't be bothered. Such wording also comes across as much more formal and insincere. Today's readers want to see your personality, your passion, your enthusiasm, instead of boring, stock phrases that are decades old.

This is the most important part for you if you want to learn how to become a better business writer – every time you put your fingers on the keyboard! I'll show you, with very practical examples, how you can become a better business writer. You'll find that effective writing can help you to achieve your goals, influence people to your viewpoint, and work more efficiently. Effective writing can also help you to be a better leader and build better relationships. This will ultimately be the key to your success.

3

21st century business language

We've discussed the importance of getting your punctuation and apostrophes in the right place and understanding how to put a sentence together. However, it's not going to make you an effective writer if you write like our great-grandfathers used to write. It might still get the job done (eventually) but using old-fashioned language will not help you stand out from the rest, and it will not help you to build relationships.

In the middle of the 20th century, business was conducted in a much more impassive, formal way than it is today. Face-to-face meetings were very formal and aloof. Similarly, the writing that evolved then became very formal, using long-winded, quite overbearing language. Writers used big words because they thought they would impress their readers, and many redundant expressions to add padding to their sentences. Passive voice was used extensively, putting a distance between the writer and the reader, which is exactly what people intended in those days.

Business today is conducted in a very informal way. In meetings and conferences we use a natural, more relaxed, friendly language rather than unnatural, formal language that was used several decades ago. We use active voice to get to the point quicker, but we still take care to be tactful. The aim in business in general is to develop relationships first, and this is done through connecting well and using appropriate language.

It really surprises me to find that, despite this, so many people are using a completely different style for their writing, and are using language in their writing that they would never use in speaking, language that is more suited to our great-grandfathers rather than to 21st century businessmen and women.

The language used in our writing today should be simple, courteous, relaxed and straightforward, quite conversational. Here are some of the key reasons why you need to relax your language in all your written communications:

1 **To establish relationships**. People get an impression of you from the first email they receive, so it's important to make a connection by using appropriate words and phrases. For example, 'We spoke' or 'As spoken' will not have the same effect on your reader as 'Thanks for your call' or 'It was great to speak to you'.

2 **To communicate your ideas precisely**. Using unsuitable or incorrect expressions, or a long-winded writing style, will not give the reader the right meaning or the right impression. It will only lead to misunderstandings, confusion and lengthy correspondence to clarify.

3 **To convey a good impression**. Clear, concise, accurate language will give an impression of efficiency, and will fill the reader with confidence. Careless or inaccurate expressions will do the opposite. Readers may question if such carelessness will extend to other business dealings too.

In this chapter we will look at the words we use to communicate, and how they are put together, so that you achieve all these aims.

 TIP Good writing is like any other endeavour. The more you put into it, the more you will get back.

Seven deadly sins of today's business writing

I conduct business writing and email etiquette workshops, and I always ask for samples of documents recently sent out by participants. Here are some of the major problems I see regularly in all types of business correspondence.

1 Redundant expressions

Expressions like 'Please be informed', 'Kindly be advised', 'I would like to bring to your attention' and 'I am writing to advise you' should have been relegated to the recycle bin way before the turn of the new millennium. Unfortunately, today's writing is still full of centuries-old expressions like these, which are simply redundant.

Instead of	Say
Please be advised that our next meeting will be held on Monday 14 June.	Our next meeting will be held on Monday 14 June.
Kindly be informed that the fire alarms will be tested at 9 am tomorrow.	The fire alarms will be tested at 9 am tomorrow.
I am writing to let you know that Mr John Lim is no longer with our company.	Mr John Lim is no longer with our company.

 CAUTION Many people think they have to use phrases like 'Please be informed' because they are more polite. You can still be courteous without using these passive, stuffy phrases.

2 Long-winded words and phrases

Our ancestors used to believe that big words and flowery sentences would impress readers. Today, such writing will only confuse or frustrate readers. Instead of *I should be very grateful*, simply say *Please* (definitely not *Kindly*). Use short words like *buy*, *try*, *start* and *end* instead of *purchase*, *endeavour*, *commence* and *terminate*. The aim should be short words, simple expressions and short sentences in short paragraphs that are clear and concise.

3 Passive voice

Our ancestors used passive voice because they didn't want to show who was responsible for anything. Passive voice also put a distance between the writer and the reader, which is what the writer wanted in those days. Today's business writing should use active voice, which is more alive, more focused, more personalised and much more interesting and clear. (You can learn more about active voice in Chapter 3.)

Instead of	Say
Arrangements have been made for a repeat order to be despatched to you immediately.	I have arranged for a repeat order to be sent to you today.
The cause of your complaint has been investigated.	I have looked into this matter.
The seminar will be conducted by Adrian Chan.	Adrian Chan will conduct the seminar.
Sales of the X101 have exceeded all expectations.	X101 sales have gone sky high.

 CAUTION Passive voice puts a distance between you and your reader. Active voice gives your writing a focus, and is much more personal and natural.

4 Yesterday's language

It beats me why so many people are using old-fashioned writing in modern emails. 'Enclosed herewith please find our catalogue for your reference and perusal', 'With reference to your above-mentioned order', 'Kindly furnish me with this information soon'. These stuffy, old-fashioned phrases were used decades ago and they have no place in today's modern writing.

Instead of	Say
We refer to your letter of 21st October 201-.	Thank you for your letter dated 21 October.
The above-mentioned goods will be despatched to you today.	These goods will be sent to you today.
Please see the below-mentioned list of items we have in stock.	These are the items we have in stock.
Kindly furnish us with this information soon.	Please let us have this information soon.
Please find enclosed herewith a copy of our new catalogue for your reference and perusal.	I'm pleased to enclose our new catalogue, and I hope you find it interesting.
Should you require any further clarification please do not hesitate to contact me.	Please give me a call on 2874722 if you have any questions.

5 Writing differently to how we speak

Many of my workshop participants say to me, 'Oh but this is what I would say. I can't write it though, can I?' I tell them, 'Yes, of course you can'. Our ancestors didn't do this, as you have seen in number 4, but today we should definitely be writing as though we are having a conversation. If you put some thought and personality and some feeling into your writing, this will ultimately lead to developing great relationships.

Instead of	Say
Your email of this morning refers . . .	Thanks for your email.
As spoken this morning . . .	Thank you for your call this morning.
As per our telecon . . .	It was good to speak to you today.
I kindly request your approval.	I hope to receive your approval.
Appreciate if you could help to process this claim at your earliest.	Please process this claim soon.
Kindly advise which course of action you would like to take and we will proceed accordingly.	Please let us know what you would like to do.
We will issue the letter to your good self early next week.	We will send you the letter early next week.
Could I please request your kind assistance in filling out the below survey for us.	I hope you will take a few minutes to complete this survey for us.

 TIP Imagine the reader is sitting in front of you. If you find yourself writing something you wouldn't say, then change it.

6 Commas instead of full stops

It's a very common error to find commas placed where they should really be full stops. If you might be guilty of this, do read my section on punctuation from page 20 to brush up on this very important tool in getting your message across effectively. Yes, punctuation matters.

Instead of	Say
Thanks for your email, it was good to hear from you.	Thanks for your email. It was good to hear from you.
Mary is responsible for this convention, she will be in touch with you soon.	Mary is responsible for this convention. She will be in touch with you soon.
Sales have been good this year, the figures are higher than last year.	Sales have been very good this year. The figures are higher than last year.
I love your ideas for this project, however, I'd like to discuss some issues with you.	I love your ideas for this project. However, I'd like to discuss some issues with you.
Her arm was injured, therefore, she could not go to work.	Her arm was injured. Therefore, she could not go to work.
I will see you next week, meanwhile please let me see some samples soon.	I will see you next week. Meanwhile, please let me see some samples soon.

 TIP Please also see Chapter 2 on Sentence Construction Matters. There is much more help there on punctuation.

7 Thank you and Regards

Why do so many people need to say 'Regards' or 'Thank you' at the end of a message? Thank you for what? For reading my email? If you have been courteous throughout your communication (and let's face it, no matter what the circumstances, your writing should always be courteous) there should be no need to keep saying 'Thank you' over and over again just because someone read your letter or email. If you have something to thank the reader for, you might consider ending with 'Thank you for your help', 'Thank you for your patience', 'Thank you for your understanding' or something similar.

As for 'Regards', I'd suggest just dropping this. Is it really necessary? We see several alternatives too, like 'Best regards', 'Warm regards'. Personally I find them all rather cold. I'd suggest putting your personality into your writing throughout your message, and then just put your name at the end. When appropriate you could close with a little nicety like 'See you soon' or 'Have a great weekend', or 'Good luck with the meeting'. Or how about 'Many thanks'? Follow this final remark with your name and it closes your message in a very friendly, relationship-building way.

TIP Put some life into your business writing by using a natural, relaxed, friendly style, as though you are having a conversation. Instead of using yesterday's jargon, use a style that is proactive, stimulating and interesting – writing that reflects your personality.

 TIP Do check out Chapter 6 on Email Etiquette, where we will look more at suitable closes for email messages.

Ten steps to brilliant business writing

In all messages it is essential to ensure correct grammar, spelling and punctuation. However, if you are to achieve results from your writing, you need more than an ability to structure sentences correctly. You need to be able to transfer thoughts and ideas from yourself to another person. To do this effectively, you must put yourself in the place of the reader and imagine how they will accept what is written in the tone used. You must anticipate the reader's needs, wishes, interests, problems, and consider the best way of dealing with the specific situation.

Unfortunately, many people are still following a rather officious style of writing introduced decades ago by government departments, local councils, banks, insurance companies, etc. This style of writing is often unfriendly and in-efficient. However, it's great to see that such organisations are now making changes and are advocating the art of plain English.

One organisation that is leading the way in this is the Plain English Campaign. Since 1979, they have been campaigning against gobbledygook, jargon and misleading public information. Anyone can access many free guides on their informative website: www.plainenglish.co.uk. According to the Plain English Campaign, 'Plain English is a message written with the reader in mind and with the right tone of voice that is clear and concise.'

 TIP Plain English is not only faster to write and read. It also enables you to get your message across more often, more easily, and in a friendly way.

So what's the secret of composing good business messages in plain English? It's really no secret at all. If you want to be understood, you must use simple language, and put your message across in a natural way, using a courteous style, and using language you would use if you were speaking.

Here are 10 ways you can become a better business writer using plain English:

1 Write naturally and sincerely

Try to show a genuine interest in your reader and his/her problems. Your message should sound sincere while written in your own style. Write naturally, as if you are having a conversation.

Instead of	Say
We have received your letter of 12 June.	Thanks for your letter of 12 June.
I have pleasure in informing you.	I am pleased to tell you.
We do not anticipate any increase in prices.	We do not expect prices to rise.
I should be grateful if you would be good enough to advise us.	Please let me know.
Please favour us with a prompt reply.	I hope to receive a prompt reply.
Please furnish me with this information soon.	Please let me have this information soon.
Please revert to us soonest.	I hope to hear from you soon.

CAUTION Try to avoid too many sentences beginning with the same word, like 'I' or 'We' or 'Please'. If you find you have done this, you need to be creative and restructure. And remember, just because you have two sentences beginning with 'Please', it's not an excuse for you to change one to the outdated 'Kindly'.

2 Be courteous and considerate

Being courteous and considerate in your writing means:

- Reply promptly to all messages – answer on the same day if possible.
- If you cannot answer immediately, write a brief note and explain why. This will create goodwill.
- Understand and respect the recipient's point of view.
- Resist the temptation to reply as if your correspondent is wrong.
- If you feel some comments are unfair, be tactful and try not to cause offence.
- Resist the temptation to reply to an offensive letter in a similar tone. Instead, answer courteously and do not lower your dignity.

Courtesy does not mean using old-fashioned expressions like 'your kind consideration' or 'your good self'. It means showing consideration for your correspondent and being empathetic – that means showing respect for your reader's feelings. Writing in a courteous style enables a request to be refused without killing all hope of future business. It allows a refusal to be made without ruining a friendship.

Many people ask for advice about how to keep their messages short and to the point without coming across as being abrupt and cold. Let's take a look at a few examples of how some messages could be improved while retaining clarity and courtesy. Notice how all the improved versions are also so much more personal and warm:

Please find below best available rate at The Westin Sydney.

Here are our best available rates at The Westin Sydney.

Kindly see attached on my department's performance for last month.

Here is my performance report for March 201-.

Your immediate feedback will be highly appreciated so we can advise the guest accordingly.

Please let me have your urgent feedback so we can let the guest know.

As per our telephone conversation, kindly find attached a copy of my credit card authorisation form.

Thank you for your call. Attached is my credit card authorisation form.

Kindly revert back with flight schedule upon confirmation and also with names of delegates, and liaise accordingly with our Martin Lim upon arrival at airport for transportation and hotel arrangements.

 Please let me know the delegates' names and their flight schedule. They will need to liaise with Martin Lim regarding hotel arrangements when they arrive.

 As spoken earlier with regards to your quotation ref XYZ123, pls advise whether possible for your good self to come down to our company on 11 June (Wed) at 9.30am for a short discussion with my senior technical manager on this issue.

 Thanks for your call today. It would be helpful if you could attend a short meeting with my senior technical manager. Is 9.30 am on Wednesday 11 June convenient for you?

3 Use appropriate tone

If your message is to achieve its purpose, the tone must be appropriate. The tone of your message reflects the spirit in which you put your message across. Even when writing a complaint or replying to one, your message can be conveyed in a way so as not to be rude or cause offence. Ignoring the need to use an appropriate tone could result in a message that sounds aggressive, tactless, curt, rude, sarcastic or offensive. This will not meet your desired objectives.

Instead of	Say
We cannot do anything about your problem.	Unfortunately, we are unable to help you on this occasion.
This problem would not have happened if you had connected the wires properly.	The problem may be resolved by connecting the wires as shown in the handbook.
Your television's guarantee is up, so you will have to pay for it to be fixed.	Your television's guarantee has ended, so unfortunately you must bear the cost of any repairs.
I am writing to complain because I was very unhappy with the way I was treated in your store today.	I was most unhappy with the standard of service I received in your store today.

You alter your tone of voice to convey messages in different ways. Much of what you say is also interpreted through non-verbal clues – eye contact, gestures, inflections of the voice, etc. This type of 'reading between the lines' is not possible with the written word. Therefore it is vital to choose your words carefully. You can be firm or friendly, persuasive or conciliatory – it depends on the impression you wish to convey. It is important to try to get the tone right because using the wrong tone could cause real offence to your reader.

Here are some expressions to avoid in your business writing:

Your failure to reply

You did not see

We must insist

You should not expect to

Your refusal to co-operate

You have ignored

This is not our fault

I can assure you

You failed to

I have received your complaint

 TIP You can learn more about tone in Chapter 16.

4 Use simple words and short sentences

Business people today have many documents to read, so a message that is direct and straight to the point, while retaining courtesy, will be appreciated. As you work on developing your writing ability, you should remember the KISS principle. KISS stands for:

Keep

It

Short and

Simple

This means instead of long or complex words, use short ones:

Instead of	Say
commence	start
regarding	about
purchase	buy
utilise	use
require	need
endeavour, attempt	try
terminate	end
state	say
expedite	hurry, speed up
advise, inform	tell
visualise	see
despatch	send
assist	help
sufficient	enough
kindly	please

KISS also means instead of long phrases, use one word where appropriate:

Instead of	Say
I should be glad if you would	Please
in spite of the fact that	despite
with regard to	about
at the present moment in time	now
conduct an investigation	investigate
in view of the fact that	as . . . because
in the event that	if
in the very near future	soon
at a later date	later
We would like to ask you to	Please

 TIP Check out the A–Z of alternative words in Appendix 2.

Avoid these phrases altogether:

I have noticed that

It has come to my attention that

I am pleased to inform you that

I am writing to let you know that

I must inform you that

Please be informed that

Please be advised that

Thanking you in anticipation

Thank you and regards

Kindest regards

5 Use modern language

Long, old-fashioned phrases add nothing to your meaning. They are likely to give a poor impression of the writer, and may even lead to confusion. A good business message will use no more words than are necessary to convey a clear and accurate message.

Instead of	Say
We are in receipt of your letter of 12 June.	Thank you for your letter of 12 June.
We have received your letter of 12 June.	Thank you for your letter of 12 June.
Enclosed herewith you will find	I enclose
Please find attached	Attached is, or I am attaching
Would you be good enough to advise me	Please let me know
Please be reminded	Please remember
the above-mentioned goods	these goods

 TIP Today's business language should be proactive, stimulating, interesting and, most of all, it should reflect your personality.

6 Use active not passive voice

'Voice' is a grammatical term that refers to whether the subject of the sentence is acting or receiving the action. Passive voice makes your messages sound stuffy and bureaucratic. Active voice makes your writing more interesting and more lively. Using active voice can considerably improve your writing style.

Check out these two examples of a similar message:

Active voice: Tim played the violin.

Here, the subject is the actor, Tim. You can almost see Tim playing the violin, totally absorbed in his music. The sentence is alive and interesting.

Passive voice: The violin was played by Tim.

Here, the subject is the violin. The action is gone. The emphasis has been moved from the subject performing the action to the subject receiving the action. It is not so easy to visualise what is happening. The sentence is dull and boring.

Here are some tips that may help you to tell when a sentence is passive:

- Watch for sentences that start with the action rather than the actor. Sentences that start with the action are often passive.
- Watch for various forms of the verb 'to be' such as – is, are, was, were, will be, have been, should be, etc. These verbs may not always indicate that the sentence is passive, but they often give you a clue.

Passive voice adds confusion in our writing, it makes sentences longer and much less lively. Passive voice was preferred by our great-grandfathers because they did not want to show any responsibility in their writing. It created a distance between the writer and the reader, which was fine for that time. Our writing today should show responsibility, and it should be more personal and natural, more focused.

Instead of	Say
The new system was developed by our staff.	Our staff developed the new system.
The cheque was presented to the charity by the Duchess of Cambridge.	The Duchess of Cambridge presented the cheque to the charity.
Your co-operation is needed in providing the following information for the directory.	Please let me have this information for the directory.

Instead of	Say
These notes should be read carefully before completing the form.	Please read these notes carefully before completing the form.
The investigation has been concluded by our client, and the paperwork has been signed.	Our client has concluded the investigation and signed the paperwork.

TIP If you want someone to do something, start your sentence with 'Please' and follow with the active verb. Don't use 'I appreciate' or 'We would appreciate', and definitely don't try starting a sentence with 'Appreciate'.

Is passive voice ever appropriate? Yes, there are some occasions when passive voice would be more appropriate.

It may be better to make a particularly important noun the subject of the sentence, giving it extra emphasis. For example, it would be better to say:

Our restaurant has been recommended by all the leading hotels in Singapore.

This emphasises 'our restaurant' rather than:

All the leading hotels in Singapore recommend our service.

When you want to place the focus on the action, not the actor. For example:

The noise was heard all over the island.

Here, the emphasis is on the noise, not the people who made the noise.

When you want to hide something or when tact is important. For example:

An unfortunate mistake was made.
This bill has not been paid.

7 Avoid nominalisations (ie using a noun instead of a verb)

There is something about writing that makes us express ourselves more formally than we would do when speaking. For example, you might chat with a colleague in the staff canteen about how you are going to realise your department's goals. But when you sit down to write a report about it, for some reason you find yourself writing about 'the realisation of our goals'.

This habit is very common. What happens is that instead of using a verb, for example, *to realise*, the writer or speaker uses the related noun, *realisation*.

Let's look at some more examples:

Verb	Nominalisation
use	the use of
avoid	the avoidance of
improve	the improvement of
implement	the implementation of
clarify	the clarification of
avoid	the avoidance of
complete	the completion of
provide	the provision of
arrange	the arrangement of
fail	the failure of

Nominalisations appear all over our writing. They lengthen our sentences and make the writing less lively, less human and more official sounding.

Instead of	Say
We ensured the motivation of staff with the introduction of Learn at Lunch sessions.	We motivated staff by introducing Learn at Lunch sessions.
His carelessness in driving caused a serious accident.	His careless driving caused a serious accident.
Newton Hospital made a decision to expand its paediatric services.	Newton Hospital decided to expand its paediatric services.
The accountant has no expectation that they will be able to meet their deadline.	The accountant does not expect them to meet their deadline.

 CAUTION Like passive verbs, too many nominalisations make writing very dull and obscure the real meaning of a sentence.

8 Use positive language

Presenting yourself as an optimist is a well-proven strategy of success. This works in writing too. Simply let the reader know what you can do and will do, rather than what you can't and won't do.

One big tip for positive writing is to avoid using 'but' wherever possible. It generally erases everything positive that came before it. The reader will just focus on the negative. Take a look at these examples. Can you see how much more positive the second sentence sounds?

> This model is very popular but it only does 35 miles per gallon.

> This model is very popular and it does 35 miles per gallon.

Take a look at these examples of positive and negative words:

Positive words	Negative words
benefit	impossible
congratulations	damaged
delighted	unable to
generous	mistake
glad	problem
proven	loss
sale	delay
save	failure
convenient	trouble
qualified	cannot
excellent	complaint
satisfactory	inconvenient
thank you	difficulty
of course	regret
pleasure	neglected
guarantee	except

 TIP Using positive words and positive phrases will enhance the tone and improve the effectiveness of your writing.

9 Use 'you' and 'we'

Even if you are writing a message to many people, write as though you are speaking to only one person. Call the reader 'you'. If this seems strange, it might help to remember that you wouldn't use words like 'the company' or 'the participant' if you were speaking to someone face to face.

Instead of	Use
Participants must bring with them . . .	Please bring with you . . .
Advice is available from . . .	You can get advice from . . .
Our company is currently restructuring . . .	We are restructuring . . .
We always tell our clients when . . .	We will always tell you when . . .

 TIP There is nothing wrong with using 'we' and 'I' in the same message, if it's appropriate.

10 Be consistent

Consistency is very important in many ways. How many spaces do you leave between different parts of a letter or a report? When I ask my workshop participants, they often don't have standards – anything goes. I recommend having standards, and putting consistency high up on your list of priorities. Also, consider how you display a date – always the same, or in different styles? Consistency says a lot about you and about your own standards.

Instead of	Say
The people attending will be John Wilson, G Turner, Mandy Harrison and Bob from Sales.	The people attending the next committee meeting will be John Wilson, Gloria Turner, Mandy Harrison and Bob Turner.
I confirm my reservation of a single room on 16/7 and a double room on 17 Oct.	I confirm my reservation of a single room on 16 July and a double room on 17 October.

What's wrong?

The following letter is full of yesterday's jargon, passive voice and long-winded writing. Can you identify everything that is wrong with it?

Dear Sirs,

We have received your letter dated 27 March 201-.

We are extremely distressed to learn that an error was made pertaining to your esteemed order. Please be informed that the cause of your complaint has been investigated and it actually appears that the error occurred in our packing section and unfortunately it was not discerned before this order was despatched to your good self.

Arrangements have been made for a repeat order to be despatched to you immediately and this should leave our warehouse later today. It is our sincere hope that you will have no cause for further complaint with this replacement order.

Once again we offer our humblest apologies for the unnecessary inconvenience that you have been caused in this instance.

Kindly contact the undersigned if you require any further clarifications.

Very truly yours,
Zachariah Creep & Partners

 TIP When you are writing, ask yourself whether you would say this if you were speaking to the recipient. Eliminate useless jargon by writing as you would speak.

Here is the same letter written in modern language.

Dear Mr Tan

YOUR ORDER NUMBER TH 2457

Thank you for your letter dated 27 March.

I am sorry to hear about the mistake made with your order. I have looked into this and the error happened in the packing section. Unfortunately, it was not discovered before the goods were sent to you.

I have arranged for a repeat order to be sent to you today, and I hope this meets your requirements.

Once again, please accept my apologies.

Please give me a call on 2358272 if you have any questions.

Yours sincerely

 TIP The key to good business writing is to write in a natural style, as if you were having a conversation.

The best writers rewrite

Rewriting is a basic principle of the best business writers, and it's a habit you must get into if you are to learn how to improve your writing. Every professional writes and then edits, and then possibly edits again. This is where the real power happens – when you cut, tighten, sharpen, refine, fix and polish.

Tightening up your writing is hard when you've had no formal training, but your goal is simple: to produce clear, concise, effective messages that connect with your readers.

 TIP Brief is in. Our digital age values short words, short sentences, short paragraphs. We want to read and understand quickly.

Short words are best

Use words with one or two syllables where possible. The fewer syllables your words contain, the faster readers will be able to read your writing. Remember, readers are impatient and have little time. Short words are also clear and understood by more people. Short words are natural, because they are used by us in our speech. People can usually trust short words more than complicated words. Don't try to impress with your knowledge of language. For today's business writing, short words are best.

Instead of	Use
attempt	try
beneficial	helpful
clarification	explanation, help
customary	usual, normal
demonstrate	show
determine	decide
modification	change
peruse	read, look at
remittance	payment
revised	new
terminate	stop, end
utilise	use

TIP Check out the A–Z of Alternative Words in Appendix 2 for many more examples of words to avoid and words to use.

Short sentences are best

Research has been conducted into the degree of understanding of sentences of different lengths. Take a look at these figures:

Number of words in the sentence	Percentage of people who will understand on the first reading
7–10 words	95%
15–20 words	75%
27 words or more	4%

Long sentences slow down reading and they will lose your reader. People are in a rush. They don't have time to read and reread to figure out what you are trying to say.

An ideal sentence length is thought to be 12 to 15 words. Sure, some sentences will need to be a little longer, and there can be no hard and fast rule, but short is always best. Here are three ways you can shorten your sentences:

- cut unnecessary words
- cut unnecessary thoughts
- break long sentences into short ones

 CAUTION Short doesn't mean abbreviations and acronyms. This may make matters worse.

Short paragraphs are best

Research has been carried on how fast people read and how much they understand. This shows that for better readability a paragraph should be around 45 words, definitely not more than 65. Three to five sentences will usually cover an individual idea. If your paragraphs are long with few white spaces, readers will need to be very motivated. So give your readers a break – literally! Break up your paragraphs by using sub-headings and numbered points where appropriate.

Short messages are best

Most readers are busy, impatient and have short attention spans. We want to know the gist of the message quickly. We want to know the point quickly. We don't want to waste time trying to work out what the reader is trying to say, and what the writer wants us to do. We expect the writer to do that.

 TIP Double your chances of getting a message read by cutting your message by half.

 # Checklist

Before signing or sending any written message, ask yourself these questions:

- Have you used simple words and simple expressions?

- Have you avoided wordiness while remembering the need for courtesy?

- Is your tone conversational and natural, as if you were speaking?

- Have you used active voice instead of passive?

- Have you used the right tone for the issue you are writing about and for the person you are addressing?

- Have you used any old-fashioned language or jargon that should be updated?

- Have you included all the essential information? Have you double-checked all the facts and figures? Is everything clear and unambiguous?

- Have you remembered consistency? For example, presentation of dates and times.

- Have you read out the letter as if speaking, to help you to tell if all the punctuation is placed correctly?

- Is your language brief but still courteous?

Structuring messages logically

Whether you are composing a business letter, a fax message, a memo or an email, the general rules for structuring the body of the message are the same. A well-structured document written in good business language is the core of effective communication. This section will help you to get past that blank page and start creating well-structured documents that will achieve your objectives.

Four-point plan

Many messages are short and routine. You can write or dictate them without any special thinking or preparation. However, documents that are not so routine need more thought and careful planning. Here is a useful, simple framework for structuring all written messages. I've been using this structure in my training for many years, and I know it works:

INTRODUCTION (Background and basics)	Why are you writing? Refer to a previous letter, contact or document.
DETAILS (Facts and figures)	Give information/instructions. Ask for information. Provide all relevant details. Separate into paragraphs. Ensure logical flow.
RESPONSE or ACTION (Conclusion)	Action the reader should take. Action you will take. Give a deadline if necessary.
CLOSE (A simple one-liner)	Sometimes all that is needed is a simple one-line closing sentence.

TIP Make your documents visually appealing and reader-friendly by using headings, numbered points and bullets where appropriate. Read more about using bullets and lists in Chapter 16.

Let's look at this four-point plan in more detail.

1 Opening or introduction

The first paragraph will state the reason for the communication, basically setting the scene. It may:

acknowledge previous correspondence

refer to a meeting or contact

provide an introduction to the matter being discussed.

> **Examples**
>
> Thank you for your letter of . . .
>
> It was good to meet you again at last week's conference.
>
> We wish to hold our annual conference at a London hotel in September.

 CAUTION You will not build relationships if you use introductions like, 'We spoke' or 'As spoken'. Say 'Thanks for your call' or 'It was great to speak to you' so that you put some feeling into your message.

2 Central section (details)

This main part of the message gives all the information that the recipient needs to know. Alternatively you may be requesting information, sometimes both. Details should be stated simply and clearly, with separate paragraphs used for individual sections. This section should flow logically to a natural conclusion, which will probably state any action needed.

3 Conclusion (action or response)

This section draws the message to a logical conclusion. It may:

state the action expected from the reader

state the action you will take as a result of the details provided.

> **Examples**
>
> Please let me have full details of the costs involved, together with some sample menus.
>
> If payment is not received within seven days, this matter will be placed in the hands of our solicitor.

4 Close

A simple one-line closing sentence is usually all that is necessary to conclude a message. This should be relevant to the content of the message.

> **Examples**
>
> I look forward to meeting you soon.
>
> I look forward to seeing you at next month's conference.
>
> I would appreciate your prompt reply.
>
> Please call me if you have any questions.

The four-point plan for structuring all written messages is illustrated in the following letter.

INSTITUTE OF SECRETARIES

Wilson House, West Street, London SW1 2AR
Telephone 020 8987 2432
Fax 020 8987 2556

LD/ST

12 May 201-

Miss Ong Lee Fong
15 Windsor Road
Manchester
M2 9GJ

Dear Lee Fong

201- SECRETARIES CONFERENCE, 8/9 OCTOBER 201-

Intro (give a brief introduction) — I have pleasure in inviting you to attend our special conference to be held at the Clifton Hotel, London on Tuesday/Wednesday 8/9 October 201-.

Details (separate paragraphs, flowing logically) — This intensive, practical conference for professional secretaries aims to:

- increase your managerial and office productivity
- improve your communication skills
- bring you up to date with the latest technology and techniques
- enable networking with other secretaries.

Leave one blank line everywhere except the signature space — The seminar is power-packed with a distinguished panel of professional speakers who will give expert advice on many useful topics. A programme is enclosed giving full details of this seminar, which I know you will not want to miss.

Conclusion (action expected from the reader) — If you would like to join us, please complete the enclosed registration form and return it to me before 30 June with your fee of £50 per person.

Close (a simple closing statement) — I look forward to seeing you again at this exciting conference.

Yours sincerely

Louise Dunscombe

LOUISE DUNSCOMBE (Mrs)

Conference Organiser

Encs

CAUTION Take care when using bullets to make sure each point is written in parallel style. Read more about constructing bullets and lists in Chapter 16.

This email message gives another example of the four-point plan.

From:	johnwang@stelectronics.co.sg
Date:	14:10:1- 12:30:45
To:	suzieliu@videoworks.com
CC:	
Subject:	24th anniversary video

Dear Suzie

Intro —— Thank you for inviting me to visit your studios last week. I was most impressed by your new facilities.

Details —— I am delighted you can accept our invitation to produce a video to celebrate the company's 25th anniversary. This is a very special landmark in our history, and it is important that this video portrays our past, present and future.

Action —— You promised to let me have a draft outlining your thoughts for this special video. I look forward to receiving this before 30 October together with your approximate costings.

Close —— If you need any further information please give me a call on 2757272.

John Wang
Marketing Manager
ST Electronics
www.stelectronics.co.sg

 TIP Study all the documents in this book as good examples of using the four-point plan. They will really help you to structure your messages logically.

Checklist

- Remember: a well-structured business document is the core of effective communication.

- Use a subject heading to give the main gist of your message.

- Refer to a previous letter, contact or document in the first paragraph – the introduction.

- Compose the central section (details) so that each point follows in a sensible order, and make sure the information flows logically from point to point.

- Separate the message into paragraphs, leaving one blank line between each section.

- Conclude your message by stating what action you expect the reader to take after reading your message.

- Be sure to include a deadline for any response, if this is appropriate.

- Your close may simply be a one-liner, whatever is relevant to the situation.

- Proofread your message carefully and take a while to consider whether it is structured appropriately and that all the details are arranged logically.

- Read through your final message as if you were the reader. Imagine how the reader will feel when receiving it. If anything is not right, make the necessary changes.

5 Presentation matters

There are many modern communication methods available today, but the traditional business letter remains an important means of sending printed messages. As the business letter acts as an ambassador for the company, it is vital that it gives a good first impression. That means it must look good, so it is helpful business practice to ensure good quality stationery and printing for the letterheaded paper. Never underestimate the importance of the 'first impression'. By setting high standards in the way you write, you will be helping to create and enhance the corporate image of your organisation.

Very often today instead of a secretary being asked to type documents for signature by the employer, it is the employer who is keying in his/her own text and sending messages straight to recipients. Although very often it makes sense in terms of time and energy for the employer to prepare his/her own communications, it is also good practice to allow a secretary to 'tighten them up'. The boss may be an expert in his/her own field or specialism, but the secretary is more often than not the expert in presentation, layout and structure.

In today's competitive business world, high communication standards are vital. Therefore, it is essential to ensure that the need for speed does not result in a decline in the standards of communication. Instead, the constant advances in technology should help us to improve and enhance our business communications, and thereby maximise business potential.

Let's take a look at this very important aspect of presentation of business documents.

Printed stationery

If you want to make a good impression when you meet someone in person, you'll make sure you look good, right? It's the same for correspondence too. If you want your documents to make a good impression, attractive and consistent presentation will certainly go a long way to help.

Your printed stationery should be of good quality, especially when being used for sending to external contacts. For internal documents the stationery does not need to be of such high quality.

The paper your company uses for its printed correspondence will express the personality of your company. Your letterhead will show:

a logo or graphic symbol identifying your company

the company's name

the full postal address

contact numbers – telephone, fax, email address

the URL or website address

registered number or registered office. When the registered office is different to that shown in the address section, it is usual to show the registered address, normally at the foot of the notepaper, along with the registered number.

It's a good idea to engage an expert to design a letterhead, especially an eye-catching logo with which the company can be identified.

Here are two examples of letterheaded paper.

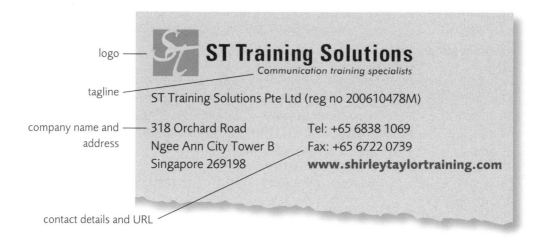

PEARSON

PEARSON EDUCATION

EDINBURGH GATE, HARLOW, ESSEX CM20 2JE
UNITED KINGDOM

TELEPHONE +44 (0) 1279 623623
FACSIMILE +44 (0) 1279 431059

www.pearson.com/uk

PEARSON EDUCATION LIMITED, REGISTERED OFFICE
EDINBURGH GATE, HARLOW, ESSEX CM20 2JE
REGISTERED NUMBER 872828
REGISTERED IN ENGLAND & WALES

Fully blocked style with open punctuation

The fully blocked layout is now the most widely used method of display for all business documents. This style has been commonly used now since the mid-1970s. It is thought to look very businesslike and sleek. This layout reduces typing time as there are no indentations for new paragraphs or the closing section. Fully blocked layout simply means that every line is aligned to the left margin. No paragraphs are indented, no headings are centred – everything starts at the left.

Open punctuation is often used with the fully blocked layout. Again this reduces typing time because there is no need for any unnecessary full stops and commas. Open punctuation means missing out all non-essential full stops and commas, for example at the end of each line of the address, after the salutation and closing section. You will learn more as you go through this chapter.

Although fully blocked layout is used by many organisations, some still prefer to adopt their own in-house style for document layout. That's fine too, as long as you remember the golden rule of consistency. It's not good practice to use one layout today and another one tomorrow, or to leave one line space between paragraphs in some letters and two in others. Be consistent. Consistency is good.

Fully blocked layout with open punctuation has been used for all the specimen documents in this book. In the business letter shown here, note the consistent spacing (only one single line space) between all sections of the letter.

Letterheaded paper	**PEARSON**
Reference (initials of writer/typist, sometimes a filing reference)	ST/PJ
Date (day, month, year)	12 November 201-
Inside address (name, title, company, full address, postal code)	Mr Alan Hill General Manager Long Printing Co Ltd 34 Wood Lane London WC1 8TJ
Salutation	Dear Alan
Heading (to give an instant idea of the theme)	FULLY BLOCKED LAYOUT
Body of letter (one line space between paragraphs)	This layout has become firmly established as the most popular way of setting out letters, fax messages, memos, reports – in fact any business document. The main feature of fully blocked style is that all lines begin at the left-hand margin.
	Open punctuation is usually used with the fully blocked layout. This means that no punctuation marks are necessary in the reference, date, inside address, salutation and closing section. Of course, essential punctuation must still be used in the text of the message itself.
	Be consistent too in layout and spacing of all documents. It is a good idea to leave just one clear line space between each section.
	I enclose some other examples of fully blocked layout as used in fax messages and reports.
	Most people agree that this layout is very attractive and easy to produce as well as businesslike.
Complimentary close	Yours sincerely
Name of sender	*Shirley Taylor*
Sender's designation or department	Shirley Taylor Author and Communication Trainer
Enc (if anything is enclosed)	Enc
Show if any copies are circulated (if more than one, use alphabetical order)	Copy: Pradeep Jethi, Publisher Amanda Long, Acquisitions Executive

CAUTION Don't leave extra spaces between paragraphs so that the spacing becomes inconsistent. This will not look good or give a good impression. Leave the same space between each paragraph and before the complimentary close.

Continuation pages

Some companies have printed continuation sheets that are used for second or subsequent pages of business letters. Such printed continuation sheets usually show just the company's name and logo. If printed continuation sheets are not available, the second or subsequent pages should be typed on plain paper of a similar quality to that of the letterhead.

When a second or subsequent page is necessary, always include certain details at the top of the continuation sheet. These details are necessary for reference purposes in case the first and subsequent pages are separated in any way:

    page number

    date

    name of addressee

When a continuation sheet is necessary, remember these guidelines:

- It is not necessary to include anything at the foot of the previous page to indicate that a further page follows. Please don't put *Cont'd* or *Continued*. It's not necessary. The fact that there is no closing section or signature should make this quite obvious.

- A continuation page should contain at least three or four lines of typing as well as the usual closing section.

- Do not leave one line of a paragraph either at the bottom of the previous page or at the top of the next page. Try to start a new page with a new paragraph.

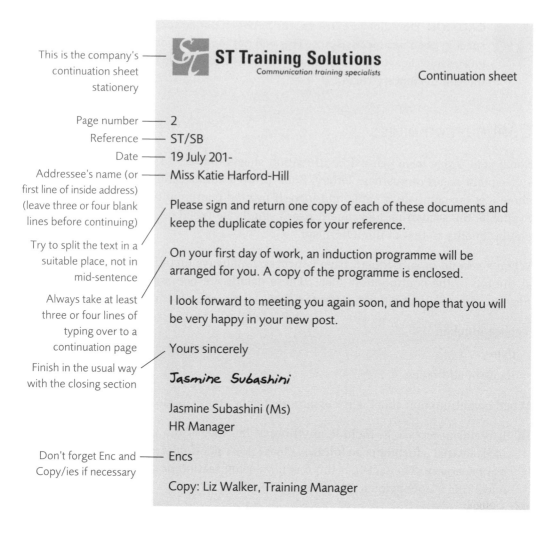

This is the company's continuation sheet stationery

ST Training Solutions
Communication training specialists
Continuation sheet

Page number —— 2

Reference —— ST/SB

Date —— 19 July 201-

Addressee's name (or first line of inside address) —— Miss Katie Harford-Hill

(leave three or four blank lines before continuing)

Try to split the text in a suitable place, not in mid-sentence

Please sign and return one copy of each of these documents and keep the duplicate copies for your reference.

On your first day of work, an induction programme will be arranged for you. A copy of the programme is enclosed.

Always take at least three or four lines of typing over to a continuation page

I look forward to meeting you again soon, and hope that you will be very happy in your new post.

Finish in the usual way with the closing section

Yours sincerely

Jasmine Subashini

Jasmine Subashini (Ms)
HR Manager

Don't forget Enc and Copy/ies if necessary —— Encs

Copy: Liz Walker, Training Manager

Parts of a business letter

1 Reference

In the past, letterheads used to have 'Our ref' and 'Your ref' printed on them. Today this is rarely the case because with modern computers and printers it is difficult to line up the printing on such pre-printed stationery. Instead, the typist normally inserts the reference on a line on its own. The reference includes the initials of the writer (usually in upper case) and the typist (in upper or lower case, as preferred). A file or departmental reference may also be included.

Examples

GBD/ST   GBD/st/Per1   GBD/ST/134

2 Date

I think it's important to show the date in full. Different countries have different conventions for displaying the date, different organisations too. Whichever style you use, remember to be consistent.

Examples

12 July 201-   July 12, 201-

3 Inside address

Display the name and address of the recipient on separate lines exactly as it would appear on an envelope. Take care to address the recipient exactly as they sign their letters. For example, a person signing as 'Douglas Cowles' should be addressed as such in the inside address, preceded by the courtesy title 'Mr'. To address him as 'Mr D Cowles' would not be appropriate.

Example

Mr Douglas Cowles
General Manager
Cowles Engineering Co Ltd
12 Bracken Hill
Manchester
M60 8AS

When writing letters overseas, the name of the country should be shown on the final line of the address section. As the letter will be sent by airmail, this should be indicated one clear line space above the inside address. Again note that the appropriate courtesy title (Mr/Mrs/Miss/Ms) should always be shown:

> **Example**
>
> AIRMAIL
>
> Mr Doug Allen
> Eagle Press Inc
> 24 South Bank
> Toronto
> Ontario
> Canada M4J 7LK

4 Special markings

If a letter is confidential it is usual to include this as part of the inside address, one clear line space above it. This may be typed in upper case or in initial capitals with underscore.

> **Example**
>
> CONFIDENTIAL
>
> Miss Iris Tan
> Personnel Director
> Soft Toys plc
> 21 Windsor Road
> Birmingham
> B2 5JT

Some decades ago an attention line was used when the writer simply wanted to ensure that the letter ended up on a certain person's desk, even though the letter was addressed to the company in general, and always began 'Dear Sirs'.

> **Example**
>
> FOR THE ATTENTION OF MR JOHN TAYLOR, SALES MANAGER
>
> Garden Supplies Ltd
> 24 Amber Street
> Sheffield
> S44 9DJ
> Dear Sirs

In today's business letters, I don't believe it should be necessary to use an attention line at all. When you know the name of the person you are writing to, include the name in the inside address, and then use a personalised salutation, ie Dear Mr Lim or Dear Leslie.

5 Salutation or greeting

If the recipient's name has been used in the inside address, it is usual to use a personal salutation.

Examples

Dear Mr Ashe   Dear Patrick   Dear Miss Farrelly   Dear Rosehannah

If your letter is addressed generally to an organisation and not to a specific person, the more formal salutation 'Dear Sirs' should be used. If your letter is addressed to a head of department or the head of an organisation whose name is not known, then it would be more appropriate to use a salutation as shown here.

Examples

Dear Sirs   Dear Sir or Madam

6 Heading

A heading gives a brief indication of the content of the letter. It is usually placed one clear line space after the salutation. Upper case is generally used for a heading on letters, although initial capitals with underscore may be used if preferred.

Example

Dear Mrs Marshall

INTERNATIONAL CONFERENCE – 24 AUGUST 201-

7 Closing section or complimentary close

In formal business letters, it has long been customary to end your letter using a complimentary close. This still stands today, and I'm not sure when or even if this may change because it is a very deep-rooted tradition. The two most common closes are 'Yours faithfully' (used only with Dear Sir/Sirs/Sir or Madam) and 'Yours sincerely' (used with personalised salutations).

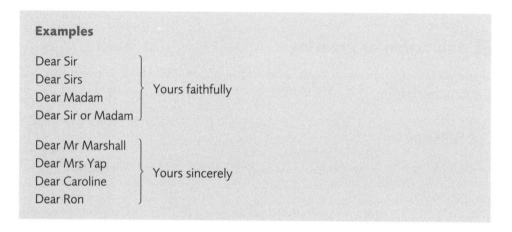

Examples

Dear Sir
Dear Sirs
Dear Madam   } Yours faithfully
Dear Sir or Madam

Dear Mr Marshall
Dear Mrs Yap
Dear Caroline   } Yours sincerely
Dear Ron

TIP We will look at emails in Chapter 6. There are different styles for the greeting and sign-off in email messages.

8 Name of sender and designation

After the complimentary close, leave four or five clear line spaces where the sender will sign the letter. The name of the sender should then be indicated in whatever style is preferred – all in upper case, or with initial capitals only. The sender's designation or department should be shown directly beneath his/her name. In these examples note that the title 'Mr' is never shown when the writer is male. However, it is a good idea to add a courtesy title for a female; this is shown in brackets after her name.

Examples

Yours faithfully	Yours sincerely
Patrick Ashe	Lesley Bolan (Mrs)
Chairman	General Manager

When the sender is not around and someone else has to sign the letter on his/her behalf, it is usual to write 'for' or 'pp' in front of the sender's printed name ('pp' is an abbreviation for *per procurationem*, which simply means 'on behalf of').

9 Enclosures

There are many different methods of indicating that something is being sent along with a letter. The first two are a little old-fashioned now, but some organisations are still known to practise these:

- Affix a coloured 'enclosure' sticker usually in the bottom left-hand corner of the letter.
- Type three dots in the left-hand margin on the line where the enclosure is mentioned in the body of the letter.
- Type 'Enc' or 'Encs' at the foot of the letter, leaving one clear line space after the sender's designation. This is the most common form of indicating enclosures.

Example

Yours sincerely

Leda Camboque (Ms)
Training Manager

Enc

10 Copies

When a copy of a letter is to be sent to a third party (usually someone in the sender's organisation) this may be indicated by typing 'cc' (copy circulated or courtesy copy) or 'Copy' followed by the name and designation of the copy recipient. If there are two or more copy recipients, it is usual to show these in alphabetical order.

Example

Copy	Ravi Gopal, General Manager
	Candice Reeves, Accountant
	Tina Choo, Marketing Director

If the writer does not wish the recipient of the letter to know that a third person is receiving a copy of the letter, then 'bcc' (blind courtesy copy) is used. This should not be shown on the top of the letter, only on the file copy and bcc copy/ies.

Example

bcc   Christian Chong, Chief Executive

 CAUTION Many managers complain about receiving too many ccs, so please don't send a cc to people just because you can. Send copies only to people who really need the information.

Open punctuation

Open punctuation is commonly used with fully blocked layout. Only punctuation marks that are essential to ensure good grammatical sense are included within the main body of the message itself. All other commas and full stops are omitted, especially in places like the date, inside address, etc.

 TIP You can refresh yourself on punctuation rules in Chapter 2.

Dates

25 September 201-   25th September, 201-
14 July 201-   14th July, 201-

Names and addresses

Mr G P Ashe	no full stops
Chief Executive	
Ashe Publications Pte Ltd	no commas at the end of lines
#03–45 Ashe Towers	
212 Holland Avenue	
Singapore 2535	

Salutation

Dear Patrick   no comma

Complimentary close

Yours sincerely   no comma

Abbreviations

Mr	eg	Mr.	e.g.
Dr	ie	Dr.	i.e.
BA	pm	B.A.	p.m.
IBM	am	I.B.M.	a.m.
MRT		M.R.T.	
NB		N.B.	
PS		P.S.	

Times and numbers

9.30 am		9.30 a.m.	9.30am
0950		0950 am	
1400		1400 hrs	
1.	8.	1	8.
2.	9.	2	9.
3.	10.	3	10.

TIP Using open punctuation will remove a lot of the clutter from your writing and presentation.

Fax messages

A fax machine is an essential item of equipment for any business. Fax messages may be sent between branches of the same company or to external business associates. When the fax was first introduced, many documents that would usually have been sent by letter were instead sent by fax. Now of course the use of email has overtaken both letters and fax messages.

When referring to the examples in this book, the text of the messages may be used in fax messages or indeed as email messages.

Printed form or template

Many companies have a standard printed form for use when sending fax messages. Very often a template is designed for calling up on computer systems. Operators need then just key in the relevant information. Here is an example of how a printed fax form or a template might be designed.

FAX MESSAGE	
To	From
Company	Date
Fax No	No of Pages (including this page)

Fully blocked style

When a pre-printed form is not available, the fully blocked style may be used in preparing a fax message, as shown in this example.

Letterheaded paper ——

Include the main —— heading 'FAX MESSAGE'

These headings are important so that all the essential details can be inserted alongside

It is important to state the number of pages being sent

A salutation may be included if preferred

The heading should state the main topic of the fax message

The body of the fax message should be composed similarly to a business letter

A complimentary close is not necessary

① Turner Communications   Mobile Phone specialists

21 Ashton Drive
Sheffield   Tel +44 114 2871122
S26 2ES   Fax +44 114 2871123
                                       Email TurnerComm@intl.uk

FAX MESSAGE

To   Susan Gingell, General Manager
Company   Asia Communication (Singapore) Pte Ltd
Fax Number   65 6767677
From   Low Chwee Leong, Managing Director
Ref   LCL/DA
Date   6 June 201-
Number of Pages   1
(including this page)

VISIT TO SINGAPORE

Thank you for your call this morning regarding my trip to Singapore next month. It's very good of you to offer to meet me at the airport and drive me to my hotel.

I will be arriving on flight SQ101 on Monday 8 July at 1830 hours. I'll be staying at the Grand Hyatt Hotel, Scotts Road.

I'm very excited about my first trip to Singapore, and look forward to meeting you.

Low Chwee Leong

TIP People will judge you on how you write – so it pays to learn how to write well!

TIP Set a high standard in all your correspondence. High standards in correspondence suggest a high standard in business generally.

Checklist

- Design an attractive letterhead with a unique logo for your company's letterheaded paper.

- Use consistent layout for all your business documents. Fully blocked style with open punctuation is the most popular.

- Leave one line space between each section of your documents. Be consistent in this aspect too.

- Include the sender's name and title in the address section – an 'attention line' is not necessary.

- If there is an appropriate heading, use it. If not, leave it out.

- Remember to indicate when something is enclosed by putting 'Enc' at the end.

- Use Copy or cc (courtesy copy) when other people receive copies.

- When a letter, memo or fax is continued onto a second page, do not type anything at the foot of the first page.

- At the head of a continuation sheet show the page number, date and name of addressee (fully blocked at the left margin).

- Your business documents reflect an impression of you and your company. Make sure it's a good one.

Electronic Communication

'Computers are magnificent tools for the realisation of our dreams, but no machine can replace the human spark of spirit, compassion, love and understanding.'

Sir Louis Gerstner

The Internet has really changed the way we work, and that also means it's changed the way we write. Let's look at email first. Email is possibly one of the greatest inventions of our lifetime. It's having a phenomenal effect on the way we communicate, but not always for the better! Reading, writing and managing email is taking an increasing amount of our time. However, research shows that the major cause of email stress is not its volume, but its inappropriate use as a communication tool.

More of us are using email to stay in touch while we are travelling or working from home. We're using not just our desktops and laptops, but also handheld devices.

We're using email to communicate with friends and family, as well as with business contacts, both at home and overseas. In our personal email, perhaps we can relax standards a little, but emailing for business has reached a new dimension. People whose jobs never used to involve writing skills are now finding themselves replying to dozens of emails every day. Most of us comment about the increasing quantity of messages we receive, and the pressure we are under to respond quickly. But when we are under such pressure, what happens to the *quality* of the messages we exchange?

Whether you are writing a thank you note, a meeting reminder, a proposal or a sales pitch, what you write and how you write it affects what people think of you, and it affects the image of your organisation.

 TIP A well-written message that looks and sounds professional will make it easier for people to want to do business with you. It will help people feel good about communicating with you. It will help you achieve the right results.

The Internet also means we've had to learn new writing rules for websites, blogs and social media like Facebook, LinkedIn and Twitter (to name just a few!). The online world is a unique medium, and writing effectively for it takes some considerable effort. Even professional writers for print publications have to adapt their style when writing for websites and blogs. In Chapter 7, I've included some simple techniques you can use for writing effectively for websites and blogs, as well as for social media sites.

Whether writing for email or social media, the bottom line remains: just as a handshake and eye contact say something about you and your organisation when you meet someone in person, the approach you take when writing anything – whether it's an email message, a blog post or a status update – will give an impression as well. You must make sure it's a good one!

6 Email etiquette

The explosive growth of email has created many problems, mainly because there have never been any strict standards or guidelines on how to use it. An informal online culture has evolved over the years, but there has never been one definitive guide to common standards and expectations among users. Consequently, systems have been overloaded, miscommunication has been rampant, reputations have been damaged, feelings have been hurt, and time has been wasted.

Another consideration is that because of the apparent informality of email, things are often written in an email message that would not be written in a business letter. This can have serious implications for some businesses, which may find themselves facing legal action.

Organisations are now realising the importance of protecting themselves against the dangers of email. However, this is not enough. They must also take steps to ensure that email works effectively for them. This chapter will take you through the good, the bad and the ugly of email, and will help you to make email work for you. You will learn how to enhance your online communication skills and create a good electronic rapport with your customers and colleagues.

 TIP Use email appropriately to enhance your reputation and your organisation's image.

Seven deadly sins of working with email

Read these statements and tick the boxes that apply to you. If you tick more than a couple of these items, you need help with your email.

☐ You sometimes wish you could backtrack after sending a message, but it's too late.

☐ You sometimes forget to add an attachment that you promised in the message.

☐ You have sometimes sent messages via email when you know a telephone call would have been better.

☐ You haven't done any housekeeping or deleted any messages for a long time.

☐ You have sent private or confidential messages via email, which you have later regretted.

☐ You sometimes send messages off quickly without a greeting or a sign-off.

☐ You often send messages without checking through for good grammar, spelling and punctuation.

 CAUTION If you have written a message in anger, leave it in your outbox for at least an hour. Then go back and look at it again. If you still feel the same way and want to send it, do so. But chances are you won't!

The good, the bad and the ugly of email

Email is really a double-edged sword, with many reasons to love and to hate it. Let's take a look:

Why do we love it?	**Why do we hate it?**
It's informal.	Has it become too informal? People are now using sms code in emails, which can't be good.
Messages can be sent to many people at the click of a button.	Do we receive too many emails just because it is so easy?

Why do we love it?	Why do we hate it?
You can attach a file and send it very easily.	If we aren't careful we may download a file that contains a virus; also large files take a longer download time.
It's instant – messages are delivered in seconds.	It can be too instant. Have you ever clicked 'send' and then wished you could retrieve it?
It's relatively cheap.	We receive lots of junk mail (spam).
It's time-zone friendly.	Lots of people send ccs and bccs just because they can – not because we need to see them.
It's always online unless you switch it off.	It means constant interruptions to your working day, so it interferes with your planned work.
You can write anything to anyone.	There is no confidentiality.
It can be prioritised. You decide which email to read first when you open your mail.	There's a real pressure to reply quickly to emails, which causes an increase in stress levels. One of the main causes of workplace stress is the pressure of keeping up with email messages.
It's easy.	Our oral communication skills are suffering. Workers are sending emails to people sitting at the next desk instead of walking over to speak to them.

 TIP Once your message is sent, it may be read by the recipient within seconds. You cannot call it back for second thoughts. So proofread it carefully before you click 'send'.

Top 10 lazy email habits

We're all busy. We all have overflowing in boxes. But lazy email habits could lead to misunderstanding, frustration, non-action, wasted effort, wasted time,

damaged relationships and ruined reputations. I often ask my workshop participants what annoys them about email they receive. Here are some of the things
they tell me. Are you guilty of any of these lazy habits? If so, it's time to take
action now, before it's too late.

1 Using a vague or outdated subject line

An email with a subject line like 'Update' or 'Hi' or 'News' is not likely to inspire
me to open it. Similarly an old subject line like 'Meeting on Tuesday' is useless
when the meeting happened last week and today's email is attaching a proposal.
People will sometimes not even read a message unless the subject line captures
their attention in the first place. Busy people with lots of messages every day
just do not have the time. Help your reader to understand the bigger picture by
composing a savvy subject line that tells the reader exactly what the message
is all about. Your subject line should be brief (many mailers will cut off long
subject lines) and should give the essence of the content of the message. If you
know the recipient gets a lot of messages, you may want to include URGENT or
FOR ACTION in the subject line of important messages.

 TIP What you put in your subject line can mean the difference between
whether your message is read right now, today, tomorrow, next week or
never.

2 Not using a greeting or sign-off

Internally, I can understand if you sometimes drop the 'Hi John' at the beginning, especially if the email exchange continues, but externally there's no
excuse. It's important to remember the simple courtesies of an appropriate greeting and sign-off. And that doesn't mean 'Thanks and regards'!

3 Not proofreading

Have you ever sent an email to the wrong person? Have you ever misspelled the
reader's name? Have you ever mentioned the wrong date for a meeting? Imagine
my embarrassment recently when I read back an email sent from my iPhone
saying: 'Hell get a $50 credit toasted our annual conference' (*Hell* instead of *He'll*,
and *toasted* instead of *towards*). I'm much more careful now when I send messages
using my phone.

CAUTION Proofread carefully, especially when emailing from your phone. If you regularly make errors in emails, people will question your attention to detail and your ability to handle your work.

4 Using abbreviations or acronyms

You may think these will save time, but they can lead to confusion for readers. While FYI is globally recognised as 'for your information', FYA could cause chaos because some people think it's 'for your action' while others think it's 'for your approval' – there is a big difference. Only use acronyms that the reader is sure to understand.

 CAUTION A word of warning with FYI. So many people complain that they often spend time trying to figure out why they received an email with FYI, when one sentence from the sender would have answered that question. Your reader can't be expected to be a mind reader.

5 Using the wrong case

Sloppy work sometimes results in writers using ALL CAPITALS or lower case abbreviations. In the world of email, using capital letters is equivalent to SHOUTING! (Didn't you automatically raise your voice then, even when reading this to yourself?) This is seen as rude and annoying, and it should not be necessary even to put a subject line in capital letters. Other people (perhaps because they are trying to mask their poor grammar skills) cannot be bothered with any capital letters at all, and this can be very frustrating, as can abbreviations like the ones seen in the message on page 90. Sometimes a message can be abbreviated so much that it can be impossible to read.

6 Not thinking about who really needs to see the message

One common aggravation of email is that people are being included in the cc list just for the sake of it, or are being forwarded copies of long messages that they don't really need to see. Some people forward messages without even a courteous note to tell the reader why it is being forwarded in the first place. And way too many people are clicking 'reply all' when only the sender needs to see the reply. The result is that everyone else is receiving a copy of a message that they don't need to read, meaning they have to open them, read them and delete them.

TIP Think carefully before you click 'reply all' and think carefully before you send ccs. Send messages to everyone who needs to know, not to everyone you know!

7 Writing everything in one long paragraph

Long-winded sentences, repetitions, lack of paragraphing or no space between paragraphs makes it clear that the writer has not tried to make his or her message reader-friendly. Some messages just ramble on and on, showing no respect for the reader's time or ease of understanding. When I receive an email that's all one huge paragraph, it's impossible to focus, to pick out the main points, to find any action items, or to respond effectively. Make it easier for your readers by structuring your messages logically and by leaving a blank line between your (short) paragraphs. For help in structuring your message logically, see Chapter 4.

8 Vague messages with details missing

If you've ever read an email and wondered what you're supposed to do, you know how frustrating this can be. People tell me they often find it difficult to figure out what action is needed. Email messages are often sent in such a rush that the writer does not plan the message beforehand and makes no effort to structure it properly. Other writers ramble on in long-winded epistles that miss the point altogether. Such messages will rarely achieve their objectives and will only serve to frustrate and confuse.

Another concern is when the sender asks you two questions and you answer only one. This will create more work for everyone, and will also cause frustration and may even damage the relationship. Before you click 'send', scan through the sender's email again to make sure you've answered all points. Make sure you include all essential information: dates, times, places, names, action points. Otherwise that inevitable 'ding-dong' will begin, wasting time and causing frustration. Again it comes back to proofreading carefully to ensure everything is included and nothing is missed.

 TIP The Internet has made it possible for us to communicate with people from all over the world. The only way those people can form an opinion of us is by looking at the way we write. Don't risk ruining your credibility with one swift click of the 'send' button.

9 Using unfriendly tone

It's very difficult to convey emotions in our writing. This often gets people into trouble because they type out exactly what they would say without thinking of the tone of voice that would be used to signal their emotions. It is so easy for misunderstandings to occur in email. If your tone is not quite right, your reader could easily be put on the defensive. Take time to read messages carefully and add some extra words if necessary. Learn more about tone in Chapter 3.

10 Messages that are just plain sloppy

As more people use email, sloppy work is becoming a bigger problem. Common complaints include: not being clear on the goal of the email, not attempting to develop a logical flow of ideas, not doing spellcheck, poor typing habits. Sending out messages with spelling, punctuation or grammatical errors is a sure way to lower your credibility. Remember that your messages give an impression of your company and of you as an individual. Make sure it is a good impression by taking as much care when composing an email message as you would with a formal letter.

 CAUTION I have one more major annoyance, and that is not receiving a reply to a business email. This means senders have to keep sending 'Did you receive this?' messages, plus it will surely damage your reputation because people will say things like, 'She never answers her email'. Take some time to make sure every email receives an appropriate response.

Top 10 tips for making the most of email

1 Turn off your email alert

Someone once told me that every time they receive a new email, their email alert shouts 'Yabadabadoo!' just like in *The Flintstones*! Whether you have a ping, a clang, a bong or a Yabadabadoo, for goodness sake turn it off occasionally. Don't let email take over your life. When you have an important project to work on, switch off your instant messaging system. Many people are now realising the value of doing this in being able to focus better, and some are checking mail just twice or three times a day.

2 Respond to messages promptly

Once you have read an email, it's common courtesy to reply promptly. If you are pushed for time and cannot reply immediately, don't feel compelled to do so. This will only result in a rushed message, perhaps mistakes, and it may not be as detailed or effective as it could be. Instead send a quick note saying you will get back to the writer soon with a more considered response. This way the writer knows you have received the message and are dealing with it.

3 Think carefully about your subject heading

This should give the recipient a good idea of the contents of the message and will make for easier handling and filing. Don't use subject lines line 'Enquiry' or 'Help needed'. These are less than useless and may not be read by a busy person scanning the subject lines quickly.

Remember the acronym SMART for your subject lines, so they are

Specific

Meaningful

Appropriate

Relevant

Thoughtful

4 Keep caps lock off

Capitals are difficult to read, they INDICATE SHOUTING and can appear THREATENING! And also don't use lots of exclamation marks!!!!! Also pls dont use lower case letters with abbreviations n acronyms. if u write in this way u r thot of as lazy and not businesslike!!!!!

5 Get your greeting and closing right

Formality doesn't read well in email. Replace formal salutations like 'Dear Pat' with informal 'Hi Pat' or even just 'Pat'. Similarly, replace 'Yours sincerely' with 'Best wishes' or even 'Cheers'. Try to avoid overuse of the very boring 'Regards', or worse still abbreviations like 'Tnks & Rgs'! Think of something novel and different if you must. Here are some nice ones I've seen recently: Over to you, To your success, Cheery greetings, Your friend, Take care, Until next time, Take it easy, Smiles.

6 Send a cc to those who need to know, not to everyone you know

If we are suffering from overflowing inboxes, how much of it is self-inflicted? Has it become too easy to send messages to lots of people just because you can? We must learn to use email more thoughtfully by recognising when we should and should not send messages. Do you really need to send all those cc, bcc and fwd copies? If you receive lots of messages that you don't really need to see, tell the authors so that it doesn't happen again.

7 Check your message carefully and get it right first time

There's no second chance with emails – once you hit 'send' it will be in the recipient's mailbox within seconds! If you don't check it carefully, it may result in the inevitable 'ding-dong' that we all know rather too well.

8 Do regular housekeeping

Delete or file your sent and read messages so as to keep your system fast and efficient. Do your filing and deleting regularly. Set up filters on your email system so that messages are automatically sent to different folders according to the sender and the subject matter. Filters will also delete unwanted messages, and some will highlight key messages with priority codes or colours.

 CAUTION Filters are great, but only when you check your folders regularly. If mail sits in the folders for ages unread, then it is defeating the objective. (You know who you are!)

9 Take pride in your message

When composing on-screen, it is easy to allow sentences to become very long. Make an effort to keep sentences short and simple, and check your syntax. The more pride you take in your message composition, the more successful you will be in being understood and achieving the desired results.

10 Don't panic – you can always pick up the phone

Have you ever been involved in a prolonged email exchange that lasts for days? Wouldn't it be better to pick up the phone? It's a sure fact that email overload is contributing to a decline in oral communication skills. People are sending emails to the person in the next office rather than walking a few steps. Remember that it's good to talk and don't let email result in the death of conversation.

 TIP When you discuss something in person or on the telephone, you can get to the root of the issue much quicker and resolve any problems so much faster.

Building relationships online

Unlike face-to-face communication, with email we can't see the writer, so we can't read any clues that may help us to interpret the message, eg tone of voice, gestures, body language. Therefore, email holds a great potential for misunderstanding and misinterpretation – as many of us have already found out when an email sent or received has clearly made the wrong impression to what we intended.

One of the main uses of email is to keep in touch with customers (and they may be internal or external), answer any queries, resolve any problems, etc. So here we will take an important look at some general principles of using email to enhance your relationships with your customers and co-workers.

 TIP Personalise each message and craft your messages around the relationship you're building with that individual.

Creating a real bond

Studies show that it takes just 15 seconds for your customer to judge you when you first meet and greet him or her. During these 15 seconds your customer will decide if they will listen to you, believe you and trust you. More important, these 15 seconds will determine if they will buy from you!

Of course, in email you don't have the benefit of body language. All you have are words, so it's important to learn how to use only your words very carefully to create your own email body language. When you learn to do this, you will be making a real connection – and that's what good customer care is all about.

 TIP Sometimes it's not *what* you say that's most important. It's *how* you say it. This is the case in email too.

Compare these two email messages and decide which one you think is going to help give a better impression and create a better bond.

1 Subject: Re: Problem solved

Hey John – We spoke this morning, and note your problem is solved. Should you require any further assistance kindly revert.

Thanks & Regards

Mary Tan

2 Subject: Re: Problem solved

Hi John

Thanks for your call today.

I'm so pleased that we've been able to find a solution to this. Good luck with future progress on this project.

I'll be here when you decide how we can help you again.

Mary

The first message was written in such a stiff, formal, unnatural way using words the writer would never use in speaking. My workshop participants actually say to me, 'Oh that's what I would say, but I can't write that'. Yes you can! You must use email to help you build relationships, not break them. You will not be doing this if you write in a different style to how you speak. You could undo all your good work on the phone by emailing in a different and more formal way.

 TIP In your emails, use a natural language, in a conversational style, just like you would if you were speaking.

Tips for creating a bond

It's not just important to get a result, to get the job done. A key aim in any communication must be to create an important connection with all your business contacts – a special bond. This applies whether you are dealing with a person face-to-face, on the telephone, in business meetings or by email.

You can create a special bond if you remember these tips:

1 **Use the customer's name.** Everyone likes to hear their name, so use it. Begin your messages with a greeting and finish off with your name.

2 **Avoid jargon.** Try not to use words that the customer may not understand. Explain clearly and simply, in everyday language.

3 **Be friendly.** You don't want to come across as apathetic or indifferent. Smile and show warmth – it will make a difference. (Yes, you can put a smile in your email messages too!)

4 **Be confident and competent.** You must come over as knowledgeable instead of hesitant and unsure. Don't beat about the bush in your email. Be clear but courteous.

5 **Show empathy.** This is not the same as agreement. You must show that you appreciate the other person's point of view or their problems, and a clear understanding of their feelings.

 TIP You could get your job done by just going through the motions without giving any extra. But by giving a little extra in your day-to-day work, your contacts will feel better and you will feel better too. These principles are not only good for business, but they prove an enjoyable and satisfying way to work.

Creating electronic rapport

Putting into practice the principles of good customer care is easier in person than on the telephone or email because you're dealing with a real person who you can see and hear. It's easier on the telephone than email because even without seeing the person you can hear them and the tone of voice used – and yes, you can hear a smile in a voice. On email you haven't got either of these advantages, so you have to take other steps to try to create electronic rapport with customers as well as colleagues. Some techniques you can use are:

1 Lead your reader into the message

Many writers tend to start their message quite irrationally, jumping straight into the topic blindly with no introduction. Ease the reader into the message by backtracking or giving some basic background information to set the scene. Be warm and friendly where appropriate.

Further to our meeting yesterday about your new project.

 It was good to meet you yesterday. It provided a good opportunity to learn more about your new project, which sounds very interesting.

 We spoke this morning regarding the above issue.

 I am glad we were able to speak on the phone this morning to clarify this issue.

 As per our teleconversation this morning.

 Thanks for calling me today. It made a nice change to speak to you instead of always dealing through email!

 We refer to your email about the progress of the above-mentioned project.

 Your news today is interesting – it sounds like you've been working really hard to ensure the success of this project.

2 Show some emotion

Some people give just the mere facts and only the facts. Others are so keen to get straight to the point that they forget to include any emotions, any feelings. Try to remember that emotive and sensory words add texture and dimension to the general message of what is being written.

You owe it to customers and colleagues to show empathy through your email, using language that will help you to form a better bond. For example:

 I'll be pleased to help you to sort out this issue.

 I appreciate your understanding in trying to resolve this problem.

 I hope I can shed some light on this soon.

I see what you mean and can appreciate your concern.

 This has shown me a clearer perspective and I can see a true picture now.

 I'm happy to offer you an extra discount of 5% in the circumstances.

 CAUTION Don't add so much emotion that you come over as too gushy. You only need an extra few words to really add a little more to your message and show some extra warmth.

3 Use a visual language

Try to paint a picture of what you are communicating. The reader will then be able to see the image that you are trying to create. Use phrases like:

 I can see what you mean.

 This is all very clear to me now.

 This will now enable us to focus on our mutual goals.

 Your suggestions look good.

 I would like to take a look at this issue from another perspective.

Why should you care what people think?

I'm often told that some people don't care what others think of them based on their emails. They feel it's not important, as long as the message gets across. But think about it, in the morning when you get up, why do you take a shower, wear freshly ironed clothes, fix your hair and your make-up? You're most probably doing it because you care about how you look.

When you send an email in all capitals, filled with abbreviations, typos and grammatical errors, the recipient will think that you haven't taken the time to correct any mistakes. If you forward emails with no comment as to why, the recipient will wonder how it's relevant to them, and feel frustrated that they have to read through the entire chain to find out.

 TIP Perception is the only reality on email. It's your choice whether you want to be perceived as educated and courteous. If you simply don't care, then that will be obvious too.

Impressions are important, and you can control what they will be. All you have to do is understand the basic rules of email etiquette, and make a tiny bit of effort to show courtesy, consideration and common sense. It won't hurt, and believe me, the benefits are many.

 TIP If you are attending a meeting or visiting a client, you would make sure you are suitably dressed. Similarly, you must make sure you present yourself appropriately online too. Does your email look good? Have you checked to make sure there are no spelling errors? Will it make a good first impression?

Sloppy email can affect careers

I was speaking to a Human Resource colleague recently and she told me that some of her staff are often rude or, more importantly, impulsive in their emails. They prefer to 'shoot out email' rather than discuss things face to face. This can not only affect workplace relationships, but it can also affect careers. Some employees seem to think they can vent their anger on the spot and then hit the 'send' button. Unfortunately, this can easily erupt into an email war.

 CAUTION What we must remember is that email sent out in haste can come back to haunt you. Apart from the possibility of landing you in court, it lowers productivity. It's also not good for morale and job satisfaction.

TIP Identify with your reader, appreciate their feelings, and use words they will understand, written in an appropriate tone.

Email @ work

Let's take a look at some emails and comment on the good and the bad points.

What's wrong with this email?

This is a poor subject line. Be specific

Hi is OK, but use a capital letter

Abbreviations like these are only suitable for mobile phone text messages, not for emails

Use capital letters

Don't abbreviate

Capitals is like shouting, and considered rude

Avoid abbreviations like this at the close

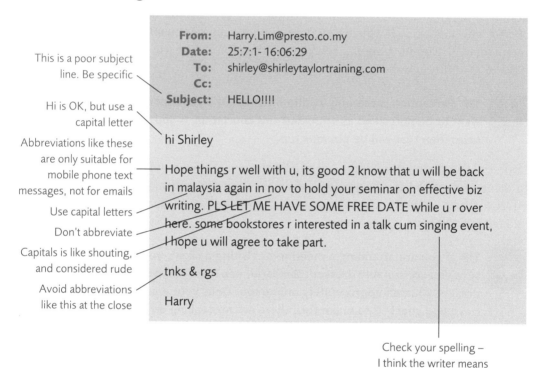

From:	Harry.Lim@presto.co.my
Date:	25:7:1- 16:06:29
To:	shirley@shirleytaylortraining.com
Cc:	
Subject:	HELLO!!!!

hi Shirley

Hope things r well with u, its good 2 know that u will be back in malaysia again in nov to hold your seminar on effective biz writing. PLS LET ME HAVE SOME FREE DATE while u r over here. some bookstores r interested in a talk cum singing event, I hope u will agree to take part.

tnks & rgs

Harry

Check your spelling – I think the writer means 'signing'!

TIP Don't use capitals in email messages. They imply SHOUTING AND AGGRESSION, they are more difficult to read, and they are not polite.

Here's the same email written more appropriately

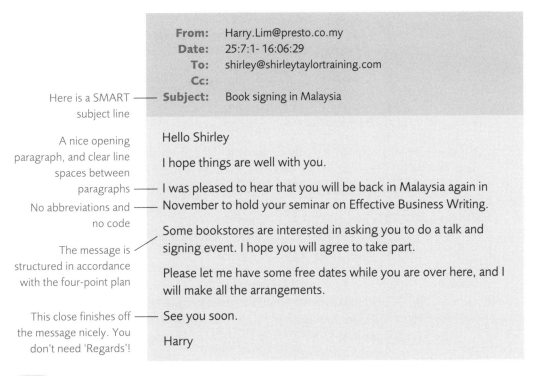

Here is a SMART subject line

A nice opening paragraph, and clear line spaces between paragraphs

No abbreviations and no code

The message is structured in accordance with the four-point plan

This close finishes off the message nicely. You don't need 'Regards'!

From: Harry.Lim@presto.co.my
Date: 25:7:1- 16:06:29
To: shirley@shirleytaylortraining.com
Cc:
Subject: Book signing in Malaysia

Hello Shirley

I hope things are well with you.

I was pleased to hear that you will be back in Malaysia again in November to hold your seminar on Effective Business Writing.

Some bookstores are interested in asking you to do a talk and signing event. I hope you will agree to take part.

Please let me have some free dates while you are over here, and I will make all the arrangements.

See you soon.

Harry

TIP Effective communication gives a professional impression of you and of your organisation. Effective communication helps to get things done.

A simple, informal email

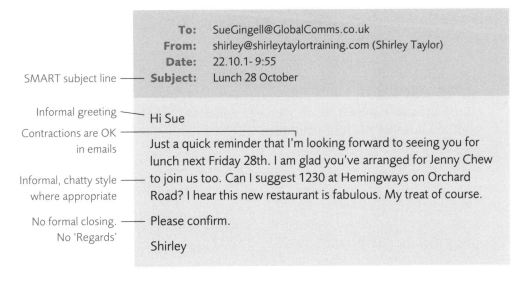

SMART subject line

Informal greeting

Contractions are OK in emails

Informal, chatty style where appropriate

No formal closing. No 'Regards'

To: SueGingell@GlobalComms.co.uk
From: shirley@shirleytaylortraining.com (Shirley Taylor)
Date: 22.10.1- 9:55
Subject: Lunch 28 October

Hi Sue

Just a quick reminder that I'm looking forward to seeing you for lunch next Friday 28th. I am glad you've arranged for Jenny Chew to join us too. Can I suggest 1230 at Hemingways on Orchard Road? I hear this new restaurant is fabulous. My treat of course.

Please confirm.

Shirley

A slightly more formal email

To:	RosehannahWethern@Pioneer.com.sg
From:	shirley@STTechnology.com (Shirley Taylor)
Date:	14.8.1- 14:30
Subject:	Customer Service Training

Dear Rosehannah

Short sentences, no 'padding' —— We are considering sending some of our staff on a training course on Customer Service Skills. Do you have a suitable course available within the next few months? If so, please let me have the dates and times plus costs.

Short paragraphs, with a space between each

If there isn't a regular Pioneer course scheduled, can you tailor-make a course specially for our staff? We could hold it in our conference room.

Write in a casual style, as if you were speaking —— Perhaps we can arrange to meet to discuss this. Are you free next Friday 20 August at 11 am? I could come over to your office if you prefer.

I hope to hear from you soon.

A standard 'signature block' —— Shirley Taylor
Project Manager
ST Technology Pte Ltd
21 Holland Lane, #01–04 Holland Court, Singapore 269198
Tel: +65 68381069 Fax: +65 67220739

Email following up a meeting

Look at how the next email message meets the requirements of the SMART subject line and the four-point plan.

From:	georgiathomas@aurorasuperstores.co.uk
Date:	10:7:1- 11:35:14
To:	lilymcbeal@healthylife.com
Cc:	richardcage@aurorasuperstores.co.uk
Subject:	Eating for Health Campaign

A SMART subject line ——

Dear Lily

Introduction —— It was good to meet you again last week. As discussed, I would like to invite you to give the opening speech at the launch of our Healthy Eating Campaign. This will be held at our Leeds superstore on Monday 8 August.

Richard and I are very excited about this campaign. We are hoping it will make the public more aware of the importance of choosing a variety of fresh fruit and vegetables as part of their daily diet.

Details —— I am attaching a provisional programme, from which you will see that 10 minutes has been allocated for the opening speech at 0930. We will be happy to arrange your transport to and from our superstore on launch day.

Action —— I know your high profile in this industry would bring the crowds flocking to this launch. We hope you will decide to join us.

Best wishes

Close ——

Georgia Thomas
Marketing Manager
Aurora Superstores Ltd
Telephone +44 114 2888724
Mobile +44 7770 2342342
www.aurora.com

 TIP When you send an email, it's your responsibility to make sure it is opened, read and acted upon. Composing a clear and specific subject line will help you to achieve this aim.

Email where tone is important

Emails are often typed and sent very quickly, without paying much thought to appropriate tone. This email is from an Administration Executive in the Accounts Department to the Manager of the Sales Department.

Read the email and consider how angry the reader would feel on reading this – angry with the sender!

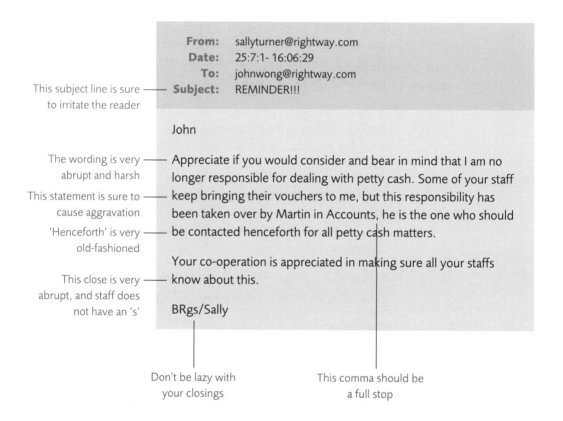

This subject line is sure to irritate the reader

From:	sallyturner@rightway.com
Date:	25:7:1- 16:06:29
To:	johnwong@rightway.com
Subject:	REMINDER!!!

John

The wording is very abrupt and harsh

This statement is sure to cause aggravation

'Henceforth' is very old-fashioned

Appreciate if you would consider and bear in mind that I am no longer responsible for dealing with petty cash. Some of your staff keep bringing their vouchers to me, but this responsibility has been taken over by Martin in Accounts, he is the one who should be contacted henceforth for all petty cash matters.

Your co-operation is appreciated in making sure all your staffs know about this.

This close is very abrupt, and staff does not have an 's'

BRgs/Sally

Don't be lazy with your closings

This comma should be a full stop

Here is the same email written in a more appropriate tone. This time the recipient would not be angry with the recipient, but with his staff.

From:	sallyturner@rightway.com
Date:	25:7:1- 16:06:29
To:	johnwong@rightway.com
Subject:	Petty Cash Vouchers

Hi John

Some of the staff from your department are still bringing their petty cash vouchers to me. However, this responsibility was taken over by Martin in Accounts last month.

Please remind your staff that they should deal with Martin in future.

Thanks for your help, John.

Sally

 TIP If you want to improve your electronic rapport with customers and colleagues, if you want to enhance your credibility and your reputation as well as your productivity, remember – it's not a computer you're talking to. It's a real live human being.

Checklist

■ Write a SMART subject line after you've written your message.

■ Include an appropriate greeting and closing section.

■ Use modern business language and simple sentences instead of old-fashioned, long-winded writing.

■ Never use ALL CAPITALS for any part of your message.

■ Learn the importance of structuring your messages logically.

■ Write as if you are having a conversation with the reader.

■ Consider the other person's feelings and make sure you use appropriate tone.

■ Format messages attractively, using full words, full sentences, and with a space between each paragraph.

■ Use email as a tool to enhance communication – not as a replacement for communication.

■ If an email exchange is getting rather long or complicated, it may be more effective to pick up the phone.

7

Writing online: websites, blogs and social networking

Good writing is even more important in social media where much of the communication is in writing. Of course, we do have videos and audio too, but for the most part readers need to be hooked in by reading the content. It's important to come across as professional, credible and likable. It will also help to consider these three Cs when writing online. Your writing must be:

Clear (readable and understandable)

Concise (straight to the point, efficient)

Consistent (reliable, dependable)

Remember too that your reader is number one! Picture them when you write; write with them in mind; write from their point of view. Ask yourself 'What's in it for my readers?' You might also want to ask yourself some questions about your readers:

- Who are they?
- What is your relationship with them?
- What do you know already?
- What do they want to know now?
- Is your language and tone appropriate?

Here are some other issues about your audience that you need to be aware of:

- They have a very short attention span.
- They have limited time.
- They want to interact rather than be passive.

■ They have a mission, a reason for reading.

■ They will more likely scan than read word-for-word.

Writing for websites

Writing for a website is different to any other kind of writing. All the rules change when it comes to writing for websites. There are two key reasons why your writing on your website must be simple, clear and concise:

1 People want to find information quickly. If you don't get to the point quickly, they will not spend a lot of time on your website.

2 People are scanning your writing to find the information they want. They are looking for keywords that attract their attention.

 TIP Not everyone will enter your site at the home page. They may follow a link to a specific page. So make sure on every page there is a clear link to your home page, a guide to pages on your site, basic information about your organisation or specialism, plus contact details.

With this in mind, here are my top 10 tips for making your website much more reader-friendly, and that means visitor-friendly.

1 Choose keywords first

It pays to pick your keywords in advance. It's much easier to tailor your writing to your keywords if you decide what they are before you start. By focusing on two or three keywords in your writing, the higher search engines will rank your website when someone searches for one of your chosen keywords.

 CAUTION When choosing keywords, try to be creative. People may type in something different or even misspellings, so it's advisable to consider including variations of the words instead of just the obvious choices.

2 Include a title on every page

As people are scanning, they really appreciate a clear headline on the top of every page of your website. This makes it easier to determine if the page contains the information they are looking for.

3 Use sub-headings

Break up your pages with sub-headings as this is again great for scanning. People aren't interested in seeing one big block of text.

4 Keep paragraphs and sentences short

Short paragraphs are much easier to read than long ones, and short sentences are also easier to read. When people are scanning websites for information, they want to find information and understand it quickly.

5 Use white space

Lots of white space will help to frame your website writing and minimise eye-strain and fatigue. The more you can do to help with this problem, the more likely it is that visitors will hang around.

6 Use bullets and numbers

Bullets and numbers help writers to keep writing brief and straight to the point. They also help readers pick out information easily.

7 Watch your language

The web is no place for long sentences and long, technical words. It's even more important when writing for a website that you keep your language plain and simple. Use simple, conversational language, as though you are having a conversation with the reader, casual and friendly.

8 Use contractions

Writing in an informal, casual style, it will come naturally to use contractions. Say 'You'll' instead of 'You will', 'It's' instead of 'It is', etc. This will help you keep sentences short and conversational.

9 Use 'sans serif' fonts like Arial

On paper, serifed fonts like Times New Roman may be easier to read. However, on your website it's important to keep your font choice simple. Arial, Tahoma, Verdana, and others that are 'sans serif' (without feet) are much easier to read on a screen.

10 Don't forget to proofread

Make sure you do everything you can to eliminate any errors from your website. Typos stand out so much more on websites, so proofread your writing, then get someone else to proofread, and do remember to use spellcheck.

 TIP The same rules as these for websites also apply to ezines, eblasts and blogs.

Writing for blogs

I have a blog, so I know how disappointed I feel when my content doesn't get the attention I think it deserves. One way we can help ourselves, apart from great promotion, is to know how to write great blog posts. Here are my suggestions:

1 Do your research

Find out what people want to know. Do your research and give them the latest information, with your own opinions thrown in too. People will respect you for giving your own thoughts.

2 Compose a killer headline

This is probably one of the keys to determining how popular your post will be. Social media users have thousands of submissions all vying to capture their attention. If you want to stand apart from the rest, you must create a compelling headline, one that is attractive enough for people to click on it.

3 Start with a great introduction

Once your reader has clicked on your killer headline, you only have a matter of seconds to convince the reader to stick around. If your opening remarks don't suck the reader in, they will be off somewhere else immediately. Your introduction should really cut to the core of what your post is about.

4 Stay focused

Online readers only have a short attention span, with no tolerance for fluff. So eliminate padding and any non-essential text. It's crucial to keep your points focused. There is a fine line between flavour and fluff!

5 Keep it 'social'

The key word in social media is 'social' – that means conversational. Write in a casual, friendly style as if you are speaking to the reader face to face. Put your personality into the content, and include readers just as you would in speaking.

6 Use the right tone

Tone comes from word and phrase choices and from use of punctuation. Consider your reader and your relationship with them. Consider how you want to be perceived. Reflect all this in the tone you use. Let your personality shine through in your writing. Let your readers really get to know the real you.

7 Avoid sales pitches

You could shoot yourself in the foot if you overpromote yourself in your blogs. First and foremost, make your content educational, informative and useful, with no ulterior motives. Avoid 'marketese'.

8 Use pictures

Pictures are great because they add a visual appeal to your post. No one likes being greeted with long blocks of text, so pictures are a great supplement for your content.

9 Be consistent

Consistency is important in the way you write, as well as your formatting. Be consistent in display, design, language use and formatting.

10 Make a commitment

Writing well for blogs is a long-term commitment, and a real learning experience. Keep trying to write great content, learn from your mistakes and from your successes. You will eventually hit on a formula that works for you.

 TIP I didn't add 'proofread' in that list, because this should be obvious. Before you hit 'send' or 'apply' or 'post', read what you wrote wearing the hat of the reader. Grammar, spelling and punctuation are important for readability as well as your reputation.

Writing for social networking sites

Not so long ago, we used to keep a diary to record our daily doings, privately. Now people update their status on social media networking sites regularly, often minute by minute. Very often though, status updates don't make compelling reading, like these for example:

Having breakfast

Walking the dog

Eating a tuna sandwich

My bus is packed today

There's no doubt, however, that social media networks are fantastic communication machines. They allow people to feel connected to a virtual community, make new friends and keep current ones, and learn things they didn't know. They encourage people to write more (which can't be bad) and to write well and concisely (which is not easy). Unfortunately, the downside is that it means people are writing instead of talking, so our face-to-face communication skills are seriously suffering.

What makes a good status update?

Most experts seem to agree that personality is what drives people to follow you, especially on Twitter, plus of course making your updates useful and interesting. If you keep posting things like 'Having coffee with Lisa', people will soon get bored with you.

So what makes a great status update? We know they need to be short, but what else? As with every written message, it's important to start by asking 'What's your aim?' or what do you want readers' reactions to be when they read your update? Perhaps it would help to consider if you want people to:

DO something. Here you are calling them to some kind of action.

THINK something. You are educating them or sharing something useful with readers.

FEEL something. You are building rapport by giving readers information that will evoke an emotion, for example it may make them smile, laugh, cry, whatever.

 TIP Remember this quote from Maya Angelou, 'People will forget what you said, people will forget what you did, but people will never forget how you made them feel.'

Crafting positive status updates

Here are three keys to writing positive status updates.

1 Be genuine

Let your personality shine through and build up rapport with your readers. Make your aim to help your fans and friends understand you better, to see more of your true personality. Avoid meaningless, mindless drivel, and TMI (too much information).

2 Be generous

It's not all about you. It's all about being helpful to your readers, your supporters, your friends. Be helpful by giving them information that will help them, including links and other resources. Social media is all about sharing.

3 Be grateful

Always show appreciation and gratitude when others are generous. A simple 'Thank you' always makes people smile, or you could share a comment, or retweet it, or repost something interesting. What goes around comes around!

 CAUTION Remember, people have short attention spans and limited time. Choose your status updates carefully, otherwise people can stop following you very quickly.

Writing techniques for social media networking sites

1 Keep posts brief

You will annoy your friends immensely if you write long-winded, unedited posts. People are scanning sites like Facebook, not reading it word for word. So keep your posts brief, and include links to longer texts or blog posts.

2 Include details

Details are important. This doesn't mean you can make your posts long (see tip 1). But if you're talking about your cat, tell us your cat's name. If you are dining with your wife, tell us her name. Giving details like this will help people feel more connected to you.

3 Share your feelings

The more feelings you share, the more authentic you will appear. If you're taking your dog to the vet, tell us how it makes you feel. If you have an important meeting tomorrow, share your feelings. This will again bring readers closer to you, helping them get to know the real you.

4 Consider your audience

Just as with any type of writing, it's important to remember who you are speaking to. All your friends will see your posts, so draw people into what you are saying by asking questions, and by revising your texts with the audience in mind.

5 Write to one person, not many

Your posts should be worded like you are speaking to one person, not to several hundreds in your network. Keep your posts conversational by using personal pronouns and contractions. Use a warm, relaxed, friendly tone.

Examples

I'm off to dinner with hubby Mark at the new restaurant Otto's in Sentosa.

I'm so excited! My book *Model Business Letters, Emails and Other Business Documents* has been translated into 8 different languages!

What are your thoughts on putting 'Regards' at the end of emails? What do you use instead?

Really looking forward to attending the Global Speakers Summit in Vancouver in December 2013. Will you be there?

---

6 Don't be vague

Vague posts are confusing and irritating. People don't want to read 'I'm about to start something exciting'. Remember the advice earlier in this chapter – clear, concise, consistent!

7 Use correct punctuation

Your post gives an impression of you, and it may be a first impression to some people. Better make sure it's a good one. Put full stops at the end of sentences, and learn the correct use of the apostrophe.

 TIP It's or Its? It's (with an apostrophe) always says either *it is* or *it has*. It never says anything else. Check out more on the apostrophe in Chapter 2.

8 Don't overuse exclamation marks

If absolutely necessary, include one exclamation mark. But please don't use multiple exclamations!!! It's very annoying!! We don't need them at the end of every sentence! OK?!!!

9 Ask questions

A great ploy to get the attention of your readers is to ask questions, or somehow draw them into a discussion. Questions make people think and respond with their own thoughts – this is the ultimate goal of any successful social media strategy.

10 Don't boast

You'll turn off your audience if you show off all the time. Don't!

11 Don't badmouth others

It's not a good idea to speak badly about people on social media. If you're mad at someone and you post about it, it will not reflect you in good light. My mum had the best advice – if you don't have anything good to say, better not say anything!

12 Maintain variety

It's not a good idea to post about the same topic all the time. People will get bored and will not respond or interact. Use an assortment of techniques – some personal comments, some links, some fun stuff, some serious stuff, some photos or videos.

A word of warning about abbreviations

So many people today are using misspellings for words, like de (the), dat (that), dis (this), wud (would), tot (thought), frens (friends), dun (don't) and even witchew (with you). This may be the language you use when texting or instant messaging friends, but it's not a good idea to get into the habit of using these non-words. When you post something on Facebook it's going out to a much wider audience. It will not give a good impression of you. In fact, it will have the opposite effect, especially if being viewed by potential employers.

If you use any abbreviations like this, do you do so all the time or just with friends? It's really important that you learn to adapt your communication style depending on who you are talking to, otherwise you could end up in trouble.

 TIP Did you know that the Romans started the practice of missing out vowels to save the cost of the messenger who usually charged per letter? This is much like what teenagers are doing today to save the cost of sending an extra text message.

CAUTION If you're not getting any comments on your posts, it should tell you something. Are your posts clear and concise? Do they show any emotion? Are they warm, positive, engaging? Do they invite interaction? The level of engagement you receive is possibly the best rating of the quality of your posts.

 Checklist

■ Be clear, concise and consistent in your online writing.

■ Keep your readers in mind, and write from their point of view.

■ Consider that your readers will have a short attention span online, and will be scanning rather than reading.

■ Use plain, simple, conversational language, as though you are having a conversation with your reader.

■ Put your true personality into your content, and write in a casual, friendly style.

■ Help your reputation by making sure your grammar, spelling and punctuation are correct.

■ Make sure status updates are useful and interesting, because personality is what will make people follow you.

■ Consider the 3 Gs of positive status updates: be genuine, generous and grateful.

■ Build rapport with your readers by considering if you want them to do something, think something or feel something.

■ Don't put anything on your social networking site that you wouldn't want a potential employer or client to see.

8 Customer care online

Most companies are realising that positive action is needed to make customer satisfaction their prime aim. If companies are to fight the increasing battle over competition, it is essential to make sure that the quality of the product or service is not only satisfactory but exceptional. The speed with which we are working these days, due to the Internet and email, has made this even more important.

In this chapter we'll take a look at this essential issue.

Why is today's customer care different?

Today's companies need to place great emphasis on marketing communications as well as providing quality customer care and aftercare so that they retain their customers in the long term. There are many more reasons why quality customer care is important today:

- increased competition
- product similarity
- better informed customers
- customers' willingness to pay for value
- rising expectations for improved support
- everyone wants everything yesterday!

The ultimate goal of successful customer care is to increase your company's market share by increasing your customers' satisfaction. All members of staff have a responsibility to help achieve this aim.

Yesterday's customer care was:	*Today's customer care is:*
Best price	Best quality
Satisfaction	Exceed expectations
Get the job done	Get the job done promptly
Competence	Relationship-building, caring

 TIP If your company is to survive in business today, your aim should be to exceed customers' expectations, even when their expectations continue to rise.

Customer care through the Internet and email

Internet use is increasing constantly. There has never been a more singular and spectacular opportunity for you to use the Internet to enhance your sales and service to your customers.

The phenomenal growth of the World Wide Web has brought about a totally new environment as far as customer relations are concerned. When a customer goes looking on the web for a specific product or service, there are literally hundreds of options to choose from. Quite simply, customers will go to the companies who have websites clearly offering the best service, the most integrated e-commerce package and exceptional after-sales service.

Let's take a brief look at what your customers expect from you in this electronic world, and how you can make the most of the Internet to maximise its potential.

Here are some general principles to follow to ensure your online sales and services are customer-friendly and web-wise:

1 Be easy to find online

Visibility is essential on the Internet, so make sure people can find you. Here are a few essentials:

- Get listings in search engines. Web marketing experts can help you here.
- Link up with other sites. Create online relationships with other sites that may have a similar audience to yours.

■   Send out a press release announcing your website in the local and national press, as well as in any trade-specific publications.

2 Ensure a visual appeal

When choosing packaging for a new product you always make sure it looks attractive and appealing. Similarly you must ensure that you present your company appropriately online. Give browsers a good first impression so that they keep coming back for more.

3 Ensure simple site navigation

People are generally impatient – they want to find things quickly and easily. You will help yourself as well as your customers by making sure site section names are clearly identified, and give clear descriptions of links.

 CAUTION Check your links carefully to make sure they are not broken.

4 Include lots of links

Include an email link on every page so that it's easy for anyone to contact you immediately.

5 Offer something extra

You can provide a great online experience by considering offering some added extras on your website, such as:

■   Links to relevant articles providing further information about your company and its products.

■   Frequently Asked Questions (FAQs) built up from experience and updated regularly.

■   Contact information and comments so that customers can send you a message easily. Include details of telephone, fax and mail address too.

■   Freebies – ask visitors to fill in a form so that they receive something for nothing. This is a good way to collect valuable data.

 TIP Put some zip into your display and formatting by getting rid of clutter.

Marketing through the web

Using advertising and direct marketing will make it easy for potential customers to find you. Some methods are:

- Give all your customers your email address and offer to answer any queries by email. Customers will appreciate this as it will avoid them making long, and sometimes expensive, phone calls.

- If phone calls are needed, use Skype. This has the added benefit of being able to see each other if you choose this option.

- Advertise your website and email addresses on your email signature, business cards, stationery, business directories and Yellow Pages, advertisements, posters, circular mailings and all promotional materials.

- Build an online community of potential customers. One of the first challenges of any email marketer is to build up a database of recipients who choose to receive email about specific topics, such as your products and services. This is called an 'opt-in' list, and it can be created by asking potential customers for their email addresses, or by including a button on your website for users to indicate their interest.

- Offer people registering for your database something for free, for example a report or e-book, or a series of several e-reports.

- Market, market, market! Once you have your electronic customer list up and running, use email as a marketing tool. Use email to send out regular information about new products, promotions, new offices, internal appointments, special announcements and newsletters. You could also consider customising the content of such mailings for each recipient by matching a user's preferences with the information you deliver.

 CAUTION Be sure to make periodic checks to ensure the information on your website is accurate and up-to-date. When information changes, you must update the details on your website.

Send out a regular e-newsletter

Email marketing is probably the most measurable, most effective direct marketing ever. There are many software programs available that will help you set up a database, create a message template and then work with you to craft an effective email campaign. All you have to do is compose the content and the software does the rest, including providing results of how many people read your message, how many people clicked on links, and much more. Email campaigns are known to be much more effective than standard direct mail, and results vary according to the strategy, frequency and professional level of the campaign, not to mention the origination of the database.

 TIP Successful e-marketing is not just about creating a website. It is about using the power of the Internet and the wonders of email to create, build and maintain prosperous and profitable customer relationships online.

Routine Business Transactions

'If you can run one business well, you can run any business well.'

Richard Branson

In any business there are always lots of routine business transactions. These include standard documents that are sent out regularly. The most common routine messages are things like enquiries, quotations and orders and their confirmations, which may be sent out on pre-printed forms. Then there are other routine dealings with customers, including payment requests, credit enquiries, follow-up letters, satisfaction surveys, and many more.

Most routine messages consist of templates that can be customised with relevant details about a customer, an issue or the item concerned. Organisations today will probably prepare routine messages using databases and mail merge technology, enabling email or printed communication to be sent efficiently and inexpensively.

Managers in charge of such templates must carefully test and review templates to ensure that they meet legal and financial requirements. It's also important that the templates are reviewed for brevity, for clarity, for errors and for unnecessary information that could get in the way of a core message. With most routine messages designed for quick scanning by the reader, writers must capture attention and convey crucial details quickly and simply.

Enquiries and replies

nquiries for information about goods or services are sent and received in business all the time. Today it's usual for routine enquiries to be made on the telephone or in a simple email. Here are some guidelines when writing an enquiry:

1. State clearly and concisely what you want – general information, a catalogue, price list, sample, quotation, etc.

2. If there is a limit to the price at which you are prepared to buy, do not mention this, otherwise the supplier may raise the quotation to the limit you state.

3. Most suppliers state their terms of payment when replying so there is no need for you to ask for them unless you are hoping for special rates.

4. Keep your enquiry brief and straight to the point.

Enquiries mean potential business, so they must be acknowledged promptly. If it is from an established customer, say how much you appreciate it. If it is from a prospective customer, say you are glad to receive it and express the hope of a lasting and friendly business relationship.

 TIP Right from the start of your business dealings, focus on building a great relationship. If you get this right first, the rest will be much easier.

Requests for catalogues and price lists

Routine requests where a formal reply is unnecessary

Suppliers receive many routine requests for catalogues and price lists. Unless the writer requests information not already included, a written reply is often not necessary, and a compliment slip may be sent instead. In the following enquiries, written replies are not necessary. The items requested may simply be mailed with a compliment slip, or of course sent in an email.

Example 1

> Dear Sir/Madam
>
> I've seen your website and am interested in your range of fax machines. Please send me a current catalogue and price list, together with copies of any descriptive leaflets that I could pass to prospective customers.
>
> Many thanks
>
> Ellie Johnson

Example 2

> Dear Sir/Madam
>
> I have seen one of your safes in the office of a local firm and they passed on your address to me.
>
> Please send me a copy of your current catalogue. I am particularly interested in safes suitable for a small office.
>
> Thanks for your help.
>
> Martin Stewart

Potentially large business

Where an enquiry suggests that large or regular orders are possible, it's important to write a personal note and take the opportunity to promote your products.

Email enquiry

Dear Sir/Madam

I have a large hardware store in Southampton and am interested in the electric heaters you are advertising in the West Country Gazette.

Please send me your catalogue and a price list.

Many thanks

Harriet Johnson (Mrs)

Reply

Dear Mrs Johnson

Thank you —— Thank you for your enquiry about our electric heaters. Please
Provide further —— take a look at our website **www.greatheaters.com**, which
information about gives full details of the products we offer. I have also put a copy
specific goods and refer of our catalogue in the mail to you today. You will find details of
to information in our prices and terms on the inside front cover of the catalogue.
catalogue

You may be particularly interested in our newest heater, the FX21 model. Without any increase in fuel consumption, it gives out 15% more heat than earlier models.

Suggest action for —— Perhaps you would consider placing a trial order to allow you
recipient to take to test its efficiency. This would also enable you to see for yourself the high quality of material and finish put into this model.

A very proactive close —— I will call you soon to answer any questions you may have.

Consider 'Best wishes' —— Best wishes
instead of 'Regards'

Requests for advice

A written reply is necessary when the enquiry suggests that the writer would welcome advice or guidance.

Email enquiry

> Dear Sir/Madam
>
> Please send me details of the catering services you provide. We often run full-day events and are looking for a collaboration with a vendor who can provide tea, coffee and refreshment snacks for breaks as well as a lunch menu.
>
> I hope to hear from you soon.
>
> John Wee

Reply

Informal greetings can be used in emails	Hi John
The introduction also refers to a courtesy call	Thanks for your email, and it was good to speak to you this morning to clarify your requirements.
Enclose the requested catalogue	I am glad I was able to talk you through our website where all our services are detailed. A pdf of our catalogue is attached, which may be helpful in making your choices.
Provide further details	Our price list is also attached, and we would be happy to offer you a 10% discount if, as you indicated on the telephone, you are able to commit to 10 events per annum.
Friendly closing, with no need for 'Regards'	If you have any questions, just give me a call.
	Sally Tan

TIP A clear style projects an efficient image, so choose your words carefully.

Enquiries through recommendations

When writing to a supplier who has been recommended, it may be to your advantage to mention the fact.

Enquiry

Dear Sirs

My neighbour, Mr W Stevens of 29 High Street, Derby, recently bought an electric lawn mower from you. He is delighted with the machine and has recommended that I contact you.

I need a similar machine, but smaller. Could you let me know what products would be most suitable for me, and send me details via mail, because I don't have email.

Yours faithfully

Reply

Dear Mrs Garson

Many thanks for your letter of 18 May. I am pleased to enclose a catalogue and price list of our lawn mowers.

The machine bought by your friend was a 38 cm *Ramsome*, which is a very efficient machine. On page 15 of our catalogue, you will see the smaller size of this model (30 cm). Alternatively, a smaller model than this is the *Panther Junior* shown on page 17.

We have both these models in stock and should be glad to show them to you if you would like to call at our showroom.

Please contact me on 2314679 to let me know when you can pop in.

Yours sincerely

Requests for samples

A request for a sample of goods provides the supplier with an excellent opportunity to present products to advantage. A reply should be convincing, giving confidence in the products.

Enquiry

Dear Sirs

We have received a number of enquiries for floor coverings suitable for use on the rough floors that seem to be a feature of much of the new building taking place in this region.

It would be helpful if you could send us samples showing your range of suitable coverings. A pattern-card of the designs in which they are supplied would also be very useful.

I look forward to hearing from you soon.

Yours faithfully

Reply

Dear Mrs King

Thank you —— Thank you for your enquiry for samples and a pattern-card of our floor coverings.

Respond to the request in the enquiry —— We have today sent to you separately a range of samples specially selected for their hard-wearing qualities. A pattern-card is enclosed.

Recommend specific samples and suggest follow-up —— For the purpose you mention, we recommend sample number 5, which is especially suitable for rough and uneven surfaces.

Enclose price list —— Our price list is enclosed, which also shows details of our conditions and terms of trading.

We encourage you to test the samples provided. When you have done this we will arrange for our technical representative to come and see you.

Give your contact number —— Please contact me on 3456891 if I can be of further help.

Appropriate close —— Yours sincerely

TIP It always helps when you consider the four-point plan for all your writing: Introduction, Details, Action and Close. For more guidance on this great way to structure your documents, see Chapter 4.

General enquiries and replies

When writing a general letter of enquiry be sure to be specific in the details required, eg prices, delivery details, terms of payment. When replying to an enquiry, be sure you have answered every specific query.

An enquiry with numbered points

When you have many points on which information is required, it may be useful to number the various points.

Enquiry

Dear Sir/Madam

Background information — During a recent visit to the Ideal Home Exhibition I saw a sample
regarding the enquiry   of your plastic tile flooring. I think this type of flooring would be suitable for the ground floor of my house, but I have not been able to find anyone in my area who is familiar with such tiling.

Would you please give me the following information:

Numbered points are — 1. What special preparation would be necessary for the
used for specific   under-flooring?
questions   2. In what colours and designs can the tiles be supplied?
     3. Are the tiles likely to be affected by rising damp?
     4. Would it be necessary to employ a specialist to lay the floor? If so, can you recommend one in my area?

Close with action — I look forward to receiving your advice on these issues.
required

Yours faithfully

Reply

Dear Mr Wilson

Thank you —— Thank you for your enquiry of 18 August regarding our plastic tile
Enclose brochure —— flooring. Our brochure is enclosed showing the designs and range
of colours in which the tiles are supplied.

Give details of local —— We work with TilesRUs, 22 The Square, Rugby, who carry
specialist out all our work in your area. I have asked their representative
to get in touch with you to inspect your floors. Someone will
call you soon to arrange a visit so they can advise you on what
preparation is necessary and whether dampness is likely to
be a concern.

Assurance of high —— Our plastic tile flooring is hard-wearing and if the tiles are laid
quality goods professionally, I am sure the work will give you lasting
satisfaction.

Please let me know if I can be of any further help.

Yours sincerely

 CAUTION Always check the original enquiry carefully to ensure you
have answered every point, and that all the information is correct.
This will avoid problems later.

First enquiries

When your enquiry is to someone you have not dealt with previously, mention
how you obtained their name and give some details about your own business.

A reply to a first enquiry should be given special attention in order to create
goodwill.

Enquiry

Background information about enquiry —

Request for details —

Further queries regarding prices for specific quantities of goods —

Dear Sir/Madam

Dekkers of Sheffield mentioned that you are manufacturers of polyester cotton bedsheets and pillow cases.

We are dealers in textiles and believe there is a promising market in our area for moderately priced goods of this kind.

Please let me have details of your various ranges including sizes, colours and prices. Samples of the different material would also be very useful.

Please state your terms of payment and discounts allowed on purchases of quantities of not less than 500 of specific items. Prices quoted should include delivery to our address shown above.

If you have any questions, please call me.

Yours faithfully

Reply by email

A personal salutation —

Thank you —
Refer to relevant publications —

Give further details regarding samples —

Reply to specific questions about quantities —

Dear Mrs Harrison

Many thanks for your enquiry of 15 January. I have today mailed our print catalogue and price list to your home address, together with a range of samples.

When you have had an opportunity to examine them, I know you will agree that the goods are excellent quality and very reasonably priced.

On regular purchases of quantities of not less than 500 individual items, we would allow a trade discount of 33%. For payment within 10 days from receipt of invoice, we will give you an extra discount of 5% of net price.

◀

Assurance of quality, —— Polyester cotton products are rapidly becoming popular because
demand and delivery   they are strong, warm and light. After studying our prices you
                           will not be surprised to learn that we are finding it difficult to
                           meet the demand. However, if you place your order not later
                           than the end of this month, we guarantee delivery within 14 days
                           of receipt.

Refer to website —— Here is a link to our **website** showing all the other products we offer.
                    If you have any questions at all, I'll be pleased to hear from you.

Friendly close —— I look forward to receiving your order soon.

Mark Farrelly

First enquiry from foreign importers

This letter is from a potential customer, so a friendly and helpful reply is necessary in order to create a good impression.

Enquiry via email

Dear Sir/Madam

We hear from Spett Mancienne of Rome that you are producing for export handmade gloves in a variety of natural leathers. I saw your website giving some details.

There is a steady demand in this country for gloves of high quality, and although sales are not particularly high, we are able to charge good prices.

Please let me know full details of the gloves you would recommend. It would also help if you can let us have samples of the various skins in which the gloves are supplied, or samples of the actual gloves.

I hope to hear from you soon.

Faizah Abdullah Rahman

Reply

Dear Mr Rahman

Thank you for your email, and I was pleased to speak to you just now about your interest in our products.

I am attaching a pdf of a more extensive catalogue of products, which I know you will find interesting. We have already arranged to courier to you samples of some of the skins we regularly use in our products. Unfortunately, we cannot send you immediately a full range of samples, but I can assure you that leathers such as chamois and doeskin, which we have not been able to send, are of the same high quality.

Mr Frank North, our European Director, will be visiting Rome early next month. I have asked him to contact you and arrange to visit you personally. He will bring with him a wide range of our goods. When you see them I know you will agree that the quality of materials used and the high standard of craftsmanship will appeal to the most selective buyer.

You also may be interested in our wide range of handmade leather handbags. They are also illustrated in the pdf attached and are of the same high quality as our gloves. Please also follow this **link** to see details online. Mr North will be able to show you samples when he calls.

Please let me know if you have any further questions.

Jonathan Turner

Requests for goods on approval

Customers often ask for goods to be sent on approval. If the customer does not want them, they must be returned within the time stated, otherwise the customer is presumed to have bought them and will be billed for them.

Customer requests goods on approval

Request

Dear Sir/Madam

Several of my customers have recently expressed an interest in your waterproof garments.

However, before placing a firm order I would like to receive a selection of men's and children's waterproof raincoats and leggings on 14 days' approval. If there are any items unsold at the end of this period, which I decide not to keep as stock, I will return them at my expense.

If quality and price are satisfactory there are prospects of very good sales here, so I hope you can give me some good news soon.

I look forward to hearing from you.

Yours faithfully

TIP Remember to use the apostrophe correctly, as in this message: *men's and children's raincoats*. Refer to Chapter 2 for more advice on the apostrophe.

Reply

Dear Mrs Turner

Thank you for enquiring about our waterproof garments.

As we have not previously done business together, I'm sure you will appreciate that I must request either the usual trade references, or the name of a bank to which we may refer. As soon as these enquiries have been made, we shall be happy to send you a good selection of the items you require.

I look forward to receiving these details from you soon, and hope that our first transaction together will be the beginning of a long and pleasant business association.

Please give me a call if you have any queries.

Name

In this reply the supplier seeks protection by asking for references. Some suppliers request a returnable deposit or a third-party guarantee. While safeguarding oneself, it is important not to offend customers by implying lack of trust.

Despatch of goods

Having received satisfactory references, the supplier sends a confident, direct and helpful letter. The reason for the low prices is given in order to dispel any suspicion the customer may have that the goods are poor quality.

Dear Mrs Turner

I have now received satisfactory references and am pleased to send you a selection of our waterproof garments.

This selection includes several new and attractive models in which the water-resistant qualities have been improved by a special process. Due to recent economies in our manufacturing methods, we have been able to reduce our prices. They are now lower than those for imported waterproof garments of similar quality.

When you have inspected these garments, please let us know which ones you wish to keep and then return the rest as soon as possible.

I hope this first selection will meet your requirements. If you would like a further selection, please let me know.

I look forward to hearing from you.

Name

Customer returns surplus

In this letter the customer informs the supplier of the goods to be kept and requests an invoice.

Dear Mrs Robinson

Thank you for sending me a selection of waterproof garments on approval.

Quality and prices are both satisfactory and I have arranged to keep the items shown on the attached statement. Please send me your invoice for £1,209.

Thank you for dealing with this very promptly.

Name

Visits by representatives

Customers often form their opinions of a company from the impressions created by its representatives. This stresses the need for careful selection and proper training of sales staff. Apart from being specialists in the art of persuasion, such representatives must also fulfil several requirements:

- They must have an excellent knowledge of the goods to be sold and the uses to which they can be put.
- They must be able to anticipate the customer's needs.
- They must be able to give sound advice and guidance to customers.
- They must aim to build relationships with all their customers.

Request for representative to call

Enquiry

> Dear Sirs
>
> I read with interest your advertisement for plastic kitchenware in the current issue of the House Furnishing Review.
>
> Can you please arrange for your representative to let me know when he is next visiting this area? I am interested to see a selection of items from your product range.
>
> This is a rapidly developing area and if prices are right I believe your products would be popular.
>
> I hope to hear from you soon.
>
> Name

Supplier's offer of visit

In this email reply the supplier uses a friendly and conversational style.

	Dear Mr Kennings
	Thanks for your email today.
It has a personal tone and does not sound like a routine reply	I am the representative for your area, and I'm happy to hear you are interested in our products. You can find full details on our comprehensive website, and I'm also attaching a pdf of our plastic goods catalogue.
It presents the case from the buyer's viewpoint — It generates interest by referring to successes	Plastic kitchenware has long been a popular feature of the modern kitchen, and its bright and attractive colours have strong appeal. Wherever dealers have arranged them in special window displays, good sales are reported.
It is proactive — It gives reasons why an order should be placed without delay	I will be in your area next week. Would it be convenient for me to call in to see you at 3 pm on Wednesday 8 October? I will bring along some samples, and am sure you will see why they are in such large demand. If you wish to have a stock of these goods before Christmas, I suggest you will need to order by the end of this month.
It is helpful and friendly	I look forward to working with you.
	Name

Requests for concessions

Customers sometimes ask for goods that are no longer available, or special terms that cannot be granted. Such requests need to be handled carefully to avoid giving offence or losing business.

Request for sole distribution rights

Enquiry

Dear Sir/Madam

Background information —— We are currently expanding our department store, and want to extend our product range.

Specific details —— We are particularly interested in your range of designer sportswear called 'Sportif', and would like to receive your trade catalogue and terms of sale and payment.

Request for sole distribution rights —— I believe your products are not yet offered by any other dealer in this town, and if we decide to introduce them we should like sole distribution rights in this area.

Suitable close —— I look forward to hearing from you.

Name

Request declined

In this reply the supplier tactfully refuses the request. The refusal is not stated in so many words but is implied in the third paragraph.

Dear Mr Sanderson

Thank you —— Thank you for your email enquiring about our Sportif designer sportswear.

Attach catalogue and give further details —— I'm attaching a pdf of the latest Sportif range, together with terms of sale and payment. We are very proud of this range of sportswear and it has been very well received, with good sales reported regularly from many areas.

A tactful response to the request for sole distribution rights —— As part of our efforts to keep down manufacturing costs, I am sure you will understand that we must increase sales by distributing through as many outlets as possible. Dealers in other areas appear to be very satisfied with their sales under this arrangement, and it appears to be working very well.

I hope we can look forward to receiving your orders soon, and would be glad to include your name in our list of approved dealers.

Express a hope for the future —— I look forward to your early reply, and do give me a call if you have any questions.

Tina Roberts

Request for special terms

Enquiry

Note the excellent use of the four-point plan in this letter.

Dear Sir or Madam

Introduction —— We are interested to receive your current catalogue and price list for your regular and mountain bikes, as well as foldable bikes.

Give details —— We are the leading bicycle dealers in Norfolk, where cycling is popular, and have branches in five neighbouring towns. If the quality of your products is satisfactory and the prices are reasonable, we expect to place regular orders for fairly large numbers.

Action required —— I hope you will allow us a special discount, as this would enable us to maintain the low selling prices that have been an important reason for the growth of our business. In return, we would be prepared to order a guaranteed annual minimum number of bicycles, the figure to be mutually agreed.

Close (with contact number) —— Please call me on 6921671 so we can discuss this personally.

Yours faithfully

Reply

In this reply the manufacturer is cautious, offering allowances on a sliding scale basis.

Dear Ms Denning

Thanks for your email and interest in our bicycles. I was sorry you weren't in your office when I called today, but I hope you received my voicemail.

I am mailing to you today a hard copy of our catalogue and price list. You can also find full details, including terms and conditions, on our website.

We have considered your suggestion of placing orders for a guaranteed minimum number of bicycles in return for a special allowance. However, after careful consideration we feel it would be better to offer you a special allowance on this sliding scale basis:

On purchases exceeding an annual total of:

£5,000 but not exceeding £8,000	5%
£8,000 but not exceeding £12,000	7%
£12,000 and above	10%

We are unable to give a special allowance on annual total purchases below £5,000.

Orders will be subject to the usual trade references.

I look forward to working with you and hope to hear from you soon. Please give me a call if you wish to discuss this. My contact details are shown in my signature below.

Best wishes

John Beaver

Letter declining special terms

In this letter a supplier tactfully refuses a request to reduce prices. Instead a counter-suggestion is made.

Dear Mr Ellis

We have carefully considered your letter of 18 December.

As we have been working together for many years, we wish we could grant your request to lower the prices of our sportswear. However, our own overheads have risen sharply in the past 12 months, and we could not reduce prices by the 15% you request unless we considerably lower our quality standards. I hope you can understand that we are not prepared to do this.

As an alternative to this 15% discount, I hope you will agree to a reduction of 5% on all our products for orders of £2,000 or more. On orders of this size we could make such a reduction without lowering standards.

I hope that you will agree to this suggestion and look forward to continuing to work with you.

Please give me a call if you have any questions.

Marsha Ellis

 TIP To develop a diplomatic writing style, imagine the person is standing in front of you and you are speaking to him/her.

Useful expressions

Requests

Opening

- We are interested in ___ as advertised recently in ___ .
- We have received an enquiry for your ___ .
- I was interested to see your advertisement for ___ .
- I understand you are manufacturers of (dealers in) ___ and should like to receive your current catalogue.

Main section and close

- When replying, please also include delivery details.
- Please also state whether you can supply the goods from stock as we need them urgently.
- If you can supply suitable goods, we may place regular orders for large quantities.

Replies to requests

Opening and main section

- ■ Thank you for your letter of ___. As requested we enclose ___.
- ■ I was pleased to learn ___ that you are interested in our ___.
- ■ Thank you for your enquiry dated ___ regarding ___.

Action and close

- ■ We look forward to receiving a trial order from you soon.
- ■ We shall be pleased to send you any further information you may need.
- ■ Please give me a call if you have any questions.
- ■ Please let me know if you need any further details.
- ■ I hope to hear from you soon.
- ■ I will call you next week to discuss this with you.

Quotations, estimates and tenders

A quotation is a promise to supply goods on the terms stated. Quotations are also sometimes referred to as bids, quotes, estimates, tenders or proposals. The prospective buyer is under no obligation to buy the goods for which a quotation is requested, and suppliers will not normally risk their reputations by quoting for goods they cannot or do not intend to supply. A quotation will be prepared on the company's letterheaded paper, and will usually include:

- an expression of thanks for the enquiry
- details of prices, discounts and terms of payment
- a clear indication of what the prices cover, eg packing, carriage, insurance
- an undertaking regarding date of delivery
- the period for which the quotation is valid
- an expression of hope that the quotation will be accepted.

Terminology

When requesting a quotation, the buyer must be careful to establish clearly whether the prices are to include such additional charges as carriage and insurance. Failure to do this may, if not specified in the supplier's quotation, lead to serious disagreement, especially where such charges are heavy, as in foreign trade dealings. Some terminology associated with quotations is shown here:

- **Carriage paid.** The quoted price includes delivery to the buyer's premises.
- **Carriage forward.** The buyer pays the delivery charges.

- **Loco, ex works, ex factory, ex warehouse.** The buyer pays all expenses of handling from the time the goods leave the factory or warehouse.
- **For (free on rail).** The quotation covers the cost of transport to the nearest railway station and of loading on to truck.
- **Fas (free alongside ship).** The quotation covers the cost of using lighters or barges to bring the goods to the ship, but not the expense of lifting the goods on board.
- **Fob (free on board).** The quotation covers the cost of loading the goods on to the ship, after which the buyer becomes responsible for all charges.
- **Ex ship.** The quoted price includes delivery over the side of the ship, either into lighters or barges or, if the ship is near enough, on to the quay.
- **cif.** Carriage, insurance and freight. Price covers charges for insurance and transport to the port named.

Routine quotations

Request for quotations for printing paper

Request

Dear Sirs

We will soon need 200 reams of good quality white poster paper suitable for auction bills and poster work generally. We are looking for paper that will retain its white appearance after pasting on walls and hoardings.

Please let us have some samples and a quotation, including delivery to our office within four weeks of our order.

Yours faithfully

This request complies with the requirements of a satisfactory letter of enquiry.

- It states clearly and concisely what is required.
- It explains what the paper is for, so it helps the supplier to quote for paper of the right quality.

- It states the quantity required, which is important because of the effect of quantity upon price.
- It states when delivery is required – an important condition in any contract for the purchase of goods.
- It states what the price is to cover – in this case 'delivery at our works'.

Quotation

The supplier's reply should be sent promptly and it should be equally business-like, ensuring that all the points from the enquiry are answered.

Dear Mr Keenan

Thank you for your enquiry dated 21 June.

We are pleased to enclose samples of different qualities of paper suitable for poster work. Here are our best prices for the various qualities, to be delivered to your office:

A1 quality Printing Paper white £2.21 per kg

A2 quality Printing Paper white £2.15 per kg

A3 quality Printing Paper white £2.10 per kg

All these papers are of good quality and quite suitable for poster work. We guarantee that they will not discolour when pasted. We can promise delivery within one week from receiving your order.

Please give me a call on 2634917 if you have any questions. We look forward to working with you.

Yours sincerely

Request for quotation for crockery

Here is another example of a request for a quotation. It states exactly what is wanted and covers the important points of discounts, packing, delivery and terms of payment.

Request

Dear Sirs

You have previously supplied us with crockery and we should be glad if you would now quote for the items listed here, manufactured by Ridgeway Pottery Company of Hanley. The pattern we require is 'number 59 Conway Spot (Green)'.

    300 Teacups and Saucers

    300 Tea Plates

    40 1-litre Teapots

When quoting prices, please include packing and delivery to our office address. Please also state discounts allowable, terms of payment and earliest possible delivery date.

I hope to hear from you soon.

Yours faithfully

Quotation

Dear Mr Clarke

CONWAY SPOT (GREEN) GILT RIMS

Thank you for your enquiry of 18 April for a further supply of our crockery. We are pleased to quote as follows:

Teacups	£83.75 per hundred
Tea Saucers	£76.00 per hundred
Tea Plates	£76.00 per hundred
Teapots, 1-litre	£4.20 each

These prices include packing and delivery, but a charge is made for crates, with an allowance for their return in good condition.

Delivery can be made from stock and we will allow you a 5% discount on items ordered in quantities of 100 or more. There would be an additional cash discount of 2% on total cost of payment within one month from date of invoice.

We hope you find these terms satisfactory. Please give me a call on 3614917 if you have any questions.

Yours sincerely

TIP Even a quotation can be structured according to the four-point plan: Introduction, Details, Action, Close. For more details of how to structure your messages logically, please see Chapter 4.

Quotations subject to conditions of acceptance

Very often a quotation is made subject to certain conditions of acceptance. These conditions vary according to the circumstances and the type of business. They may relate to a stated time within which the quotation must be accepted, or to goods of which supplies are limited and cannot be repeated. The supplier must make it clear when quoting for goods in limited supply or subject to their being available when the order is received. Examples of qualifying statements are:

- This offer is made subject to the goods being available when the order is received.
- This offer is subject to acceptance within 7 days.
- The prices quoted will apply only to orders received on or before 31 March.
- Goods ordered from our 201- catalogue can be supplied only while stocks last.
- For acceptance within 14 days.

Foreign buyer's request for quotation

Email enquiry from overseas buyer

> Dear Sirs
>
> We have recently received a number of requests for your lightweight raincoats. We are hoping to place regular orders with you, as long as your prices are competitive.
>
> From the description in your catalogue, we feel that your AQUATITE range would be most suitable for this region. Please let me have a quotation for men's and women's coats in both small and medium sizes, delivered cif Alexandria.
>
> If your prices are right, we will place a first order for 400 raincoats, namely 100 of each of the 4 qualities. Shipment would be required within 4 weeks of order.
>
> I look forward to a prompt reply.
>
> Many thanks

Quotation

The reply by the English manufacturer is a good example of the modern style in business writing. The tone is friendly and the language is simple and clear. The writer shows an awareness of the problems of the tropical resident (eg the reference to condensation) and gives information that is likely to bring about a sale (eg mention of 'repeat orders' and 'specially treated').

Another point of interest here is the statement of freight and insurance charges separate from the cost of the goods. This is convenient for calculating the trade discount and also tells the buyer exactly what is to be paid for the goods themselves. Note also the statement 'For acceptance within one month'. Here the supplier promises to sell goods at the quoted price within a given period of time.

The supplier's attempt to interest the customer in other products is very good business technique.

Subject: Aquatite Rainwear

Dear Mrs Barden

Thank you —— Thank you for your email. I am pleased to learn about the enquiries you have received for our raincoats.

Discuss popularity of product with particular reference to tropical climates —— Our AQUATITE range is particularly suitable for warm climates. During the past year we have supplied this range to dealers in several tropical countries. We have already received repeat orders from many of those dealers. This range is popular not only because of its light weight but also because the material used has been specially treated to prevent excessive condensation on the inside surface.

Mentioning repeat orders gives assurance of quality

We are pleased to quote as follows:

100 AQUATITE coats	men's	medium	£17.50 ea	1,750.00
100 AQUATITE coats	men's	small	£16.80 ea	1,680.00
100 AQUATITE coats	women's	medium	£16.00 ea	1,600.00
100 AQUATITE coats	women's	small	£15.40 ea	1,540.00
				6,570.00

Specific details regarding prices ——

less 33.33% trade discount	2,187.81
Net price	4,382.19
Freight (London to Alexandria)	186.00
Insurance	122.50
TOTAL	4,690.69

Details about terms, —— Terms: 2.5% one month from date of invoice
shipment and Shipment: Within 3–4 weeks of receiving order
acceptance For acceptance within one month.

Refer to other products —— We feel sure you may also be interested in some of our other
and enclose products. Please take a look at the attached pdf, which you may
literature print to hand out to your customers. You can also see full details
on our website.

We hope to receive your order soon, and if you have any
questions just drop me a note.

Best wishes

Laura Harrison

Tabulated quotations

Many quotations are either tabulated or prepared on special forms. Such tabulated quotations are:

- clear, since information is presented in a form that is readily understood
- complete, since essential information is unlikely to be omitted.

Tabulated quotations are particularly suitable where there are many items. Like quotations on specially prepared forms, they should be sent with a covering letter which:

- expresses thanks for the enquiry
- makes favourable comments about the goods themselves
- draws attention to other products likely to interest the buyer
- gives a link to the website
- expresses hope of receiving an order
- takes the proactive approach of saying the writer will follow up soon.

 TIP When you give the potential customer everything they need, this will create a favourable impression and help to build goodwill.

Covering letter with quotation on specially prepared form

Covering email or letter

> Dear Mrs Greenway
>
> Thank you for your recent enquiry. Our quotation for leather shoes and handbags is attached, and all items can be delivered from stock.
>
> These goods are made from very best quality leather and can be supplied in a range of designs and colours wide enough to meet the requirements of a fashionable trade such as yours.
>
> Please visit our **website** where you will find details of all our other products, including leather purses, gloves and other accessories.
>
> I will follow up with you in a few days to answer any questions. Meanwhile, please give me a call on 9635117 if anything is not clear.
>
> I look forward to working with you.
>
> Matthew Marsden

 TIP Learn how to use your email program efficiently so that you can put hyperlinks in documents like this one.

Quotation

In this quotation, note the following points:

- It is given a serial number to help in future reference.
- Using catalogue numbers identifies items with precision and avoids misunderstandings. Individual shapes and sizes are also given their own serial numbers.
- 'For acceptance within 21 days' protects the supplier in case the buyer orders goods after this time when prices may have risen.
- '4% one month' indicates that a discount of 4% will be allowed on quoted prices if payment is made within one month. For payment made after one month but within two months, the discount is reduced to 2%.

CENTRAL LEATHERCRAFT LTD

85–87 Cheapside, London EC2V 6AA
Telephone 020-7242-2177/8

Quotation no JBS/234   Date 20 August 201-

Smith Jenkins & Co
15 Holme Avenue
Sheffield
S6 2LW

Catalogue Number	Item	Quantity	Unit Price
S 25	Men's Box Calf Shoes (brown)	12 pairs	65.75
	Men's Box Calf Shoes (black)	36 pairs	65.50
S 27	Ladies' Glace Kid Tie Shoes (various colours)	48 pairs	64.80
S 42	Ladies' Calf Colt Court Shoes	24 pairs	64.35
H 212	Ladies' Handbags – Emperor	36	66.50
H 221	Ladies' Handbags – Paladin	36	78.75
H 229	Ladies' Handbags – Aristocrat	12	80.00
	FOR ACCEPTANCE WITHIN 21 DAYS		
	Delivery: ex works		
	Terms: 4% one month 2.5% two months		

(signed)

for Central Leathercraft Ltd

TIP Good writers learn to present the positives rather than the negatives.

Estimates and specifications

While a quotation is an offer to sell goods at a specific price and under stated conditions, an estimate is an offer to do certain work for a stated price, usually on the basis of a specification. Like a quotation, an estimate is not legally binding, so the person making it is not bound to accept any order that may be placed against it.

Estimate for installation of central heating

Enquiry

In this enquiry the writer encloses a specification giving a detailed description of the work to be done and materials to be used. This will provide the basis for the contractor's estimate. The plan would consist of a rough sketch (drawn to scale) showing the required positions of the radiators.

Dear Sirs

Please let me have an estimate for installing central heating in my bungalow at 1 Margate Road, St Annes-on-Sea. A plan of the bungalow is attached showing required positions and sizes of radiators, together with a specification showing further details and materials to be used.

I am interested only in first-class workmanship and I want only the best quality materials to be used. However, cost is also important. It is essential that this work is completed by 31 August.

I hope to receive your prompt reply, which should include a firm completion date.

Yours faithfully

Specification

SPECIFICATION FOR INSTALLING CENTRAL HEATING
at 1 MARGATE ROAD, ST ANNES-ON-SEA

1. Installation of the latest small-bored central heating, to be carried out with best quality copper piping of 15 mm bore, fitted with 'Ryajand' electric pump of fully adequate power and lagged under floor to prevent loss of heat.
2. Existing boiler to be replaced by a Glow-worm No. 52 automatic gas-fired boiler, rated at 15.2 kW and complete with gas governor, flame failure safety device and boiler water thermostat.
3. Installation of a Randall No. 103 clock controller to give automatic operation of the central heating system at predetermined times.
4. Existing hot-water cylinder to be replaced by a calorifier-type cylinder suitable for supplying domestic hot water separately from the central heating system.
5. Seven 'Dimplex' or similar flat-type radiators to be fitted under windows of five rooms, and in hall and kitchen, according to plan enclosed; also a towel rail in bathroom. Sizes of radiators and towel rail to be as specified in plan attached.
6. Each radiator to be separately controlled, swivelled for cleaning and painted pale cream with red-lead undercoating.
7. The system to be provided with the necessary fall for emptying and to prevent air-locks.
8. All work to be carried out from under floor to avoid cutting or lifting floor boards, which are tongued and grooved.
9. Insulation of roof with 80 mm fibreglass.

JOHN HARRIS

5 July 201-

Contractor's estimate

The contractor can calculate costs from the information provided, and will send an estimate with a covering letter or email, which should provide the following information:

- A reference regarding satisfactory work carried out elsewhere, which will give the customer confidence
- A promised completion date
- A market prices and wages adjustment clause to protect the contractor from unforeseen increases that may raise costs and reduce profits
- A hope that the estimate will be accepted.

In this letter, note that the contractor aims to inspire confidence by referring to work done elsewhere, and promises to arrange an inspection if required.

Dear Mr Harris

INSTALLATION OF CENTRAL HEATING
AT 1 MARGATE ROAD, ST ANNES-ON-SEA

Thank you ——— Thank you for your letter of 6 July enclosing specification and plan for a gas-fired central heating system.

Mention price and ——— We will be pleased to carry out this work for a total of £4,062.50
discount with a 2.5% discount if payment is made within one month of
Promised completion ——— completion. We can promise to complete all work by 31 August if
date we receive your instructions by the end of this month. Please bear
This clause protects the ——— in mind that our prices are based on present costs of materials and
contractor from labour. If these costs rise, we will have to add the increased costs
unforeseen to our price.
increases

We have installed many similar heating systems in your area. Our
Mention of satisfactory ——— reputation for first-class work is well known. If you would like to
work carried out inspect one of our recent installations before making a firm
elsewhere will give decision, I would be happy to make arrangements.
confidence

Please give me a call to discuss this and answer any questions.
I look forward to working with you on this project.

Yours sincerely

TIP Notice the closes like 'I look forward to working with you'. These will help to develop the relationship, which should always be a key aim in business. Find out more about closes in Chapter 4.

Tenders

A tender is usually made in response to a published advertisement. It is an offer for the supply of specified goods or the performance of specified work at prices and under conditions set out in the tender. A tender becomes legally binding only when it is accepted; up to that time it may be withdrawn. It is usual for tenders to be made on the advertiser's own form, which includes a specification where necessary and sets out the terms in full detail.

A public invitation to tender

THE COUNTY COUNCIL OF LANCASHIRE COUNTY HALL, PRESTON PR1 2RL

Tenders are invited for the supply to the Council's power station at Bamford, during the year 201-, of approximately 2,000 tonnes of best quality furnace coke, delivered in quantities as required. Tenders must be submitted on the official form obtainable from County Hall to reach the Clerk of the Council not later than 12 noon on Friday 30 June.

The Council does not bind itself to accept the lowest, or any, of the tenders submitted.

BRIAN BRADEN

Clerk to the Council

Contractor's letter enclosing tender

Contractors must obtain and complete the official form, and then send it back with a formal covering letter.

CONFIDENTIAL

Clerk to the Council
County Hall
Preston
PR1 2RL

Dear Mr Braden

TENDER FOR FURNACE COKE

Having read the terms and conditions in your official form, I enclose my tender for the supply of coke to the Bamford power station during 201-.

I hope to learn that this is accepted.

Yours sincerely

A closed invitation to tender

An invitation to tender restricted to members of a particular organisation or group is called a 'closed tender'. This example is taken from the *Baghdad Observer*.

STATE ORGANISATION FOR ENGINEERING INDUSTRIES
PO BOX 3093 BAGHDAD IRAQ
TENDER NO 1977
FOR THE SUPPLY OF 16,145 TONNES
OF
ALUMINIUM AND ALUMINIUM ALLOY INGOTS,
BILLETS AND SLABS

1. The SOEI invites tenderers who are registered in the Chamber of Commerce and hold a Certificate of Income Tax of this year, as well as a certificate issued by the Registrar of Commercial Agencies confirming that he is licensed by the Director General of Registration and Supervision of Companies, to participate in this tender. General terms and conditions together with specifications and quantities sheets can be obtained from the Planning and Financial Control Department at the 3rd floor of this Organisation against payment of one Iraqi Dinar for each copy.
2. All offers are to be put in the tender box of this Organisation, Commercial Affairs Department, 4th floor, marked with the name and number of the tender at or before 1200 hours on Saturday 31 January 201-.
3. Offers should be accompanied by preliminary guarantee issued by the Rafidain Bank, equal to not less than 5 per cent of the C & F value of the offer.
4. Any offer submitted after the closing date of the tender, or which does not comply with the above terms, will not be accepted.
5. This Organisation does not bind itself to accept the lowest or any other offer.
6. Foreign companies who have no local agents in Iraq shall be exempted from the conditions stated in item number 1 above.

ALI AL-HAMDANI (ENGINEER)
PRESIDENT

Quotations not accepted or amended

When a buyer rejects a quotation or other offer, it is courteous to write and thank the supplier for their trouble and explain the reason for rejection. The letter of rejection should:

- thank the supplier for their offer
- express regret at inability to accept
- state reasons for non-acceptance
- if appropriate, make a counter-offer
- suggest that there may be other opportunities to do business together.

Buyer rejects supplier's quotation

Dear Mr Walton

Thank you for your quotation dated 19 February for strawboards.

I appreciate your trouble in this matter but as your prices are very much higher than those I have been quoted by other dealers, I am unable to place an order.

I will contact you again when I require other products.

Yours sincerely

Supplier grants request for better terms

Enquiry

Dear Ms Altieri

Thank you — Thank you for your letter of 18 August and for the samples of cotton lingerie you sent to me.

Mention good quality but express concern at high prices leaving small profit — I recognise the good quality of these garments, but I feel your prices are on the high side, even for such high quality. If I accept your prices I would be left with only a small profit on my sales, since this is an area in which the principal demand is for articles in the medium price range.

Repeat feelings regarding quality and desire to do business. Request special allowance — I would like to work with you and introduce your goods to my customers. Please let me know if you are able to reduce your quoted prices.

I hope to hear from you soon.

Yours sincerely

Reply

Dear Mr Daniels

Acknowledge letter ——— Thank you for your letter of 21 May, and I'm glad you like the quality of our products. We do our best to keep prices as low as

Respond to query ——— possible without sacrificing quality, and we are constantly
regarding high prices investigating new methods of manufacture.

Give assurance that ——— Considering the quality of our goods, we feel the prices we
prices are reasonable quoted are quite reasonable. However, bearing in mind the special character of your trade, we are prepared to offer you a

Special discount on ——— discount of 4% on a first order for £2,000. We are happy to make
first order will be this revised offer to begin working with you, and hope you will
appreciated by new accept this.
customer

End on a positive note ——— I look forward to receiving your first order soon.

Yours sincerely

Follow-up letters

When a buyer has asked for a quotation but does not place an order or even acknowledge the quotation, it is natural for the supplier to wonder why. A keen supplier will arrange for a representative to call, or send a follow-up message if the enquiry is from a distance.

Supplier's follow-up letter

Here is an effective follow-up letter written in a tone that shows the supplier genuinely wants to help and in a style that is direct and straight to the point. It considers the buyer's convenience by offering a choice of action, and closes with a reassuring promise of service.

Dear Mrs Larkin

As we have not heard from you since we sent you our catalogue of filing systems, I wonder whether you require any further information before placing an order.

The modern system of lateral filing has important space-saving advantages wherever economy of space is important. However, if space is not a challenge for you, our flat-top suspended system may be suitable. This system gives a neat and tidy appearance to the filing drawers, and files can be located easily and quickly. Many users find these two features particularly beneficial.

Would you like us to send our representative to call and discuss your requirements in more detail? John Robinson, our representative in your area, has advised on equipment for many large, modern offices. He would be able to recommend the system most suitable for you. There would of course be no obligation. Alternatively, you may prefer to visit our showroom and see for yourself how the different filing systems work.

I look forward to hearing from you soon, or you may contact me on 2356123 with any questions.

Yours sincerely

CAUTION Take care with punctuation. Note in the third paragraph there is a comma after 'John Robinson' and another comma after 'our representative in your area'. For more help with commas, see Chapter 2.

Letter to save a lost customer

No successful business can afford to lose its regular customers. Periodical checks must be carried out to identify those customers who have not placed recent orders, and suitable follow-up letters must be sent.

Dear

We notice that it is some time since we last received an order from you. We hope this is in no way due to dissatisfaction with our service or with the quality of goods. In either case we would like to hear from you.

We are most anxious to ensure that customers obtain maximum satisfaction from their dealings with us. If lack of orders from you is due to changes in the type of goods you handle, we may still be able to meet your needs if you will let us know in what directions your policy has changed.

As we have not heard otherwise, we assume that you are still selling the same range of sports goods, so a copy of our latest illustrated catalogue is enclosed. We feel this compares favourably in range, quality and price with the catalogues of other manufacturers. You will see in our catalogue that our terms are now much better than previously, following the withdrawal of exchange control and other official measures since we last did business together.

I hope to hear from you soon.

Yours faithfully

Useful expressions

Requests for quotations, estimates, etc

Openings
- Please quote for the supply of ___.
- Please send me a quotation for the supply of ___.
- We wish to have the following work carried out.

Closes
- As the matter is urgent we should like this information by the end of this week.
- If you can give us a competitive quotation, we expect to place a large order.
- If your prices compare favourably with those of other suppliers, we shall send you an early order.

Replies to requests for quotations, etc

Openings

- Thank you for your letter of ___.
- We thank you for your enquiry of ___ and are pleased to quote as follows:
- We are sorry to learn that you find our quotation of ___ too high.

Closes

- We hope you will find our quotation satisfactory and look forward to receiving your order.
- We shall be pleased to receive your order.
- As the prices quoted are exceptionally low and likely to rise, please place your order without delay.
- As our stocks of these goods are limited, we suggest you place an order immediately.

Orders and their fulfilment

Placing orders

Printed order forms

Most companies have official printed order forms (see Figure 11.1). The advantages are:

■ such forms are pre-numbered and therefore reference is easy
■ printed headings ensure that no information will be omitted.

Printed on the back of some forms are general conditions under which orders are placed. It is usual to refer on the front to these conditions, otherwise the supplier is not legally bound by them.

Letter orders

Smaller companies may not have printed forms but instead place orders in the form of a letter or email. When sending an order by letter or email, always ensure accuracy and clarity by including:

■ an accurate and full description of goods required
■ catalogue numbers
■ quantities
■ prices
■ delivery requirements (place, date, mode of transport, whether the order will be carriage paid or carriage forward, etc) and
■ term of payment previously agreed.

J B SIMPSON & CO LTD

18 Deansgate, Sheffield S11 2BR
Telephone: 0114 234234
Fax: 0114 234235

Order no 237

Date 7 July 201-

Nylon Fabrics Ltd
18 Brazenose Street
Manchester
M60 8AS

Please supply:

Quantity	Item(s)	Catalogue Number	Price*
25	Pura Bed Sheets (106 cm) blue	75	£11.50 each
25	Pura Bed Sheets (120 cm) primrose	82	£11.00 each
50	Pura Pillow Cases blue	117	£6.90 each
50	Pura Pillow Cases primrose	121	£6.90 each
		(signed)	
		for J B Simpson & Co Ltd	

FIGURE 11.1 An order form

Legal position of the parties

According to English law, the buyer's order is only an offer to buy. The arrangement is not legally binding until the supplier has accepted the offer. After that both parties are legally bound to honour their agreement.

The buyer's obligations

When a binding agreement comes into force, the buyer is required by law to:

- accept the goods supplied as long as they comply with the terms of the order
- pay for the goods at the time of delivery or within the period specified by the supplier
- check the goods as soon as possible (failure to give prompt notice of faults to the supplier will be taken as acceptance of the goods).

The supplier's obligations

The supplier is required by law to:

- deliver the goods exactly as ordered at the agreed time
- guarantee the goods to be free from faults of which the buyer could not be aware at the time of purchase.

If faulty goods are delivered, the buyer can demand either a reduction in price, a replacement of the goods or cancellation of the order. It may also be possible to claim damages.

Routine orders

Routine orders may be short and formal but they must include essential details describing the goods, as well as delivery and terms of payment. Where two or more items are included on an order, they should be listed separately for ease of reference.

Confirmation of telephone order

Dear

Thanks for your call this morning. I confirm our order as follows:

10 Excelda Table Lamps, 50 cm high	Price: £45 each
5 Excelda Standard Lamps	Price: £105 each

Colour: Gold

Terms: 40% trade discount

We need these lamps urgently, so I'm glad you can arrange for immediate delivery from stock.

Please confirm.

Many thanks

 TIP Notice how the introduction begins with 'Thanks for your call this morning'. Please don't use stale expressions like 'We spoke' or 'As per our telecon' or 'As spoken'. Use modern language as if you were speaking. See Chapter 3 for more help.

Tabulated order

Dear Martin

Here is our order for books on our usual discount terms of 25% off published prices:

Number of copies	Title	Author	Published
50	Communication for Business	Shirley Taylor	£9.95
100	Email Etiquette	Shirley Taylor	£7.95

Please let me know when we can expect delivery.

Many thanks

Order based on quotation

Dear John

Thank you for your quotation of 4 June. Please supply:

Code ST3134

500 CD/DVD Sleeves white window (in packs of 100)

Price: £2.10 per 100 (including VAT)

Please deliver by the end of this month.

Many thanks

Covering email with order form

When a covering email is sent with an order form (as shown in Figure 11.1), all essential details will be shown on the form and any additional explanations in the covering email.

Dear

Thank you for your quotation of 5 July. Our order number TW-237 for four of the items is attached.

I know you have all these items in stock, and hope you can deliver them within the next few days. Please call me on 6923848 if you need to discuss.

Many thanks

Acknowledging orders

An order should be acknowledged immediately, especially if it cannot be fulfilled straight away. For small routine orders a standard email may be enough, but a short note stating when delivery may be expected also helps to create goodwill. If the goods cannot be supplied at all, you must explain why and offer suitable substitutes if they are available.

Formal acknowledgement of routine order

Dear

Thank you for your email and order number TW-237 for bed coverings.

All the items are in stock, so we will be delivering them to you tomorrow afternoon.

We look forward to receiving further orders.

Many thanks

Acknowledgement of a first order

When an order is received from a brand new customer, it should most certainly be acknowledged by phone call, email or preferably both.

Refer to the phone call; make it friendly, as if you are speaking ——	Dear Laura It was good to speak to you today, and I'm very pleased to receive your order for cotton prints.
Confirm prices and —— delivery information Give assurance of —— satisfaction	I can confirm supply of the prints at the prices mentioned in my earlier email. Delivery should be made by our own transport early next week. We feel confident that you will be completely satisfied with these goods and that you will find them of exceptional value for money.
Mention other goods —— and attach any pdfs	I wonder if you have checked out the wide range of goods we have available on our website? I'm also attaching a pdf of our cotton collection that may particularly interest you.
Close with a wish for —— future business dealings	I look forward to working with you again and to a happy working relationship. Best wishes

Acknowledgement of order pointing out delayed delivery

When goods ordered cannot be delivered immediately, an email should apologise for the delay and give an explanation. A delivery date should also be given, if possible, and express the hope that the customer is not inconvenienced.

Reason for delay: breakdown in production

Dear

Thank you for your order dated 15 March for Imron Hairdriers. I'm sorry that we cannot supply them immediately due to a recent fire in our factory.

We are hoping to resume production within the next few days and we fully expect to be able to deliver these hairdriers by the end of next week, ie 14 October.

Please confirm that we can go ahead with this new delivery date.

Once again, many apologies for the inconvenience.

Best wishes

CAUTION Notice in paragraph one the writer says 'I'm sorry that'. Please don't use 'We regret that'. This is very old-fashioned now. For more on 21st century business language please see Chapter 3.

Reason for delay: stocks not available

Dear

Many thanks for your order dated 20 January.

Unfortunately, we are out of stock of the Pink Oversized Tunic (web ID 23421). This is due to the prolonged cold weather, which has increased demand considerably. The manufacturers have, however, promised us a further supply by the end of this month. If you can wait until then we will send you the goods immediately.

Please let us have your instructions.

Name

Declining orders

There may be times when a supplier will not accept a buyer's order. Some reasons may be:

- He is not satisfied with the buyer's terms and conditions.
- The buyer's credit is suspect.
- The goods are not available.

 TIP You must take utmost care when writing to reject an order so that goodwill and future business are not affected.

Supplier refuses price reduction

When a supplier cannot grant a request for a lower price, reasons should be given.

Dear

We have carefully considered your request to reduce our price quoted for woollen lingerie. However, we are unable to change our quotation for these goods.

The prices we quoted leave us with only the smallest of margins. They are in fact lower than those of our competitors for goods of similar quality.

The wool used in the manufacture of our Thermaline range undergoes a special patented process that prevents shrinkage and increases durability. The fact that we are the largest suppliers of woollen lingerie in this country is in itself evidence of the good value of our products.

We hope you will reconsider ordering our goods. If you then still feel you cannot accept our offer, we hope it will not prevent you from contacting us again in the future.

We look forward to hearing from you again soon.

Best wishes

Supplier rejects buyer's delivery terms

When delivery terms cannot be met, the supplier should show a genuine desire to help customers in difficulty.

Dear Mr Johnson

Thank you — Thank you for your order R345 24 for 10 x ST32LW Full LED TV.
Mention that delivery date cannot be met — However, we are not able to deliver before the end of November as you requested.

Further details regarding demand for the goods and how orders are being dealt with — The manufacturers of these goods are finding it impossible to meet current demand for this popular model. We placed an order for 100 sets one month ago and were informed that all orders were being met in strict rotation. Unfortunately, we will not receive the goods before the end of January.

Give an alternative

This suggestion that the customer should try another supplier is sure to be appreciated. This will help to build goodwill — I understand from our telephone conversation this morning that your customers are unwilling to consider other brands, and I can appreciate the popularity of the ST brand. We do have reasonable stocks of the ST44LW and the ST37LW, if you wish us to supply these instead. If not, can I suggest that you try Prime Vision of Leeds. They usually carry large stocks and may be able to help you.

All the best

Name

Supplier refuses to extend credit

If a previous account remains unpaid, the utmost tact is necessary when reject-ing another order. Nothing is more likely to offend a customer than the sugges-tion that they may not be trustworthy. In this email, the writer tactfully avoids suggestion of mistrust, and instead gives internal difficulties as the reason for refusing further credit.

Dear Mr Richardson

We were pleased to receive your order BW435 for a further supply of Coffee Grinders and Bread Makers.

At present the balance of your account stands at £1,800.54, and unfortunately we are unable to grant credit for further supplies until this balance is reduced. I hope you can appreciate that we do have commitments of our own to meet, and we can only do this by asking our customers to keep their accounts within reasonable limits.

If you can send us your cheque for at least £1,000, we will be happy to supply these new goods immediately.

Thank you for your understanding.

Name

Counter-offers from suppliers

When a supplier receives an order that cannot be met for some reason, several options are available:

1 Send a substitute. However, careful judgement will be required, since there is the risk that the customer may be annoyed to receive something different from what was ordered. It is advisable to send a substitute only if a customer is well known or if there is a clear need for urgency. Such substitutes should be sent 'on approval', with the supplier accepting responsibility for carriage charges both ways.

2 Make a counter-offer.

3 Decline the order.

Supplier sends a substitute article

Dear

Many thanks for your email and order RS980.

All the items are in stock except for the 25 Macy cushion covers in strawberry pink. Stocks of these have recently sold out, and the manufacturers inform us that it will be another four weeks before they can send replacements.

As you state that you need these items urgently, we have substituted 25 Macy cushion covers in a fuschia pink. These are identical in design and quality to those ordered. They are attractive and rich-looking, and very popular with our other customers. We hope you will find them satisfactory. If not, please return them at our expense. We shall be glad either to exchange them or to arrange credit.

All items will be on our delivery schedule tomorrow. We hope you will be pleased with them.

Many thanks

TIP Notice in the first paragraph 'Many thanks for your email and order RS980'. This is better than 'We have received your order RS980'.

Supplier makes a counter-offer

If the supplier decides to make a counter-offer, a great deal of skill is needed to bring about a sale. The supplier is, after all, offering something that has not been asked for. Therefore, it is important that the suggested substitute is at least as good as the one ordered.

Dear

Thank you —— Thank you for your email and order for 800 metres of 100 cm wide watered silk.

Respond to the enquiry and apologise that the fabric ordered is no longer available —— I am sorry to say that we can no longer supply this silk. Fashions constantly change, and in recent years the demand for watered silks has fallen to such an extent that we no longer produce them.

Mention a replacement and give assurance of quality and reliability —— In their place we can offer our new Butterfly brand of rayon. This is a finely woven, hard-wearing, non-creasable material with a very attractive shine. We receive a lot of repeat orders from leading distributors and dress manufacturers, giving clear evidence of the widespread popularity of this brand. At the low

Include price information —— price of only £4.20 per metre, this rayon is much cheaper than silk and its appearance is just as attractive.

Mention other products you have available —— We also manufacture other fabrics and you can find full details on our website. All these fabrics are selling very well in many countries and can be supplied from stock. If you decide to place

Give delivery details —— an order we can arrange for shipment within one week.

Please drop me a note or call me on 6283842 if you have any questions.

I hope to hear from you soon.

Packing and despatch

When goods are despatched, the buyer should be informed either by an advice note or by email stating what has been sent, when it was sent, and the means of transport used. The customer then knows that the goods are on the way and can make arrangements to receive them.

Request for forwarding instructions

Dear

We are pleased to confirm that the 10 Carmen CY430 projectors that you ordered on 15 October are now ready for despatch.

When placing your order you stressed the importance of prompt delivery, and I am glad to say that by making a special effort we have been able to improve by a few days on the delivery date agreed.

Please let us have your shipping instructions, and as soon as we hear from you we will send you our advice of despatch.

Many thanks

Advice of goods ready for despatch

Dear

We are pleased to confirm that the books you ordered in your order XY239 dated 3 April are packed and ready for despatch. The consignment consists of two cases, each weighing about 100 kg.

We have made arrangements for shipment, cif Singapore, with Watsons plc, our forwarding agents. When we receive their statement of charges, we will arrange for shipping documents to be sent to you through RBS Bank against our draft for acceptance, as agreed.

We look forward to further business with you.

 TIP You can see commonly used terminology like 'cif' in Chapter 10.

Notification of goods despatched

Dear

ORDER NUMBER S 274

The 40 Bukhara silk rugs you ordered on 5 January have been packed in four special waterproof-lined cases. They will be collected tomorrow for consignment by passenger train, and should reach you by Friday.

We feel sure you will find the consignment supports our claim to sell the best rugs of their kind.

We look forward to receiving further orders from you.

Please call us on 26919847 if you have any queries.

Report of damage in transit

It is the legal duty of the buyer to collect any purchases from the supplier. Unless the terms of the sale include delivery, the railway or other carrier is considered the agent of the buyer. The buyer is, therefore, responsible for any loss, damage or delay which may affect the goods after the carrier has taken over.

Dear

ORDER NUMBER S 274

I am sorry to inform you that one of the four cases containing 40 Bukhara silk rugs despatched on 28 January was delivered damaged. The waterproof lining was badly torn and it will be necessary to send these 10 rugs for cleaning before we can offer them for sale.

Please arrange to send replacements immediately and charge them to our account.

We realise that the responsibility for damage is ours and have already taken up the matter of compensation with the railway authorities.

I hope to hear from you soon.

Report of non-delivery of goods

When goods do not arrive as promised, avoid the tendency to blame the supplier as it may not be their fault. Your email or letter should be restricted to a statement of the facts and a request for information.

Dear

ORDER NUMBER S 524

You wrote to us on 28 January informing us that the 40 Bukhara silk rugs under this order were being despatched.

We expected these goods a week ago and had promised immediate delivery to several customers. We naturally feel our customers have been let down.

Please find out what has happened to the consignment and let us know when we may expect delivery.

Many thanks

 CAUTION You must take particular care to watch your tone in written messages so that you do not offend or anger the reader. Find out more about appropriate tone in Chapter 16.

Complaint to carrier concerning non-delivery

Upon receiving the report of non-delivery the supplier should at once take up the matter with the carriers. The message should be confined to the facts and ask for an immediate enquiry into the circumstances.

Dear

We understand that a consignment of 40 Bukhara silk rugs addressed to W Hart & Co, 25–27 Marshall Avenue, Warrington has not yet reached them.

These cases were collected by your carrier on 28 January for consignment by passenger train, and they should have been delivered by 1 February. We have your carrier's receipt number 3542.

As our customer needs these goods urgently, please let us know the cause of the delay and when delivery will be made.

Please treat this matter as one of extreme urgency.

Many thanks

Useful expressions

Placing orders

Openings
- Thank you for your quotation of ___.
- We have received your quotation of ___ and enclose our official order form.
- Please supply the following items as soon as possible:

Closes
- As these goods are needed urgently, we look forward to prompt delivery.
- Please acknowledge receipt of this order and confirm that you will be able to deliver by ___.
- We hope to receive your advice of delivery within the next two days.

Acknowledging orders

Openings
- Thank you for your order dated ___.
- We are sorry to inform you that the goods ordered on ___ cannot be supplied.

Closes

- We hope the goods reach you safely and that you will be pleased with them.
- We hope you will find the goods satisfactory and look forward to receiving your further orders.
- We are pleased to say that these goods have been despatched today (will be despatched in ___/are now awaiting collection at ___).

Invoicing and settlement of accounts

The final stage in a business transaction is to send an invoice and receive payment of the amount owing for goods supplied or services rendered. In the retail trade, eg when you walk into a shop, supermarket or department store, transactions are usually for cash. In other business transactions, like wholesale and foreign trade, it is customary to allow credit.

Invoices and adjustments

When goods are supplied on credit, the supplier sends an invoice to the buyer to:

■ inform the buyer of the amount due

■ enable the buyer to check the goods delivered

■ enable entry in the buyer's purchases day book.

When an invoice is received it should be checked carefully, not only against the goods supplied but also for the accuracy of both prices and calculations. Depending on your location, you have rights as either the supplier or the receiver on goods sent between companies. If you are unsure you should seek advice from the relevant local body.

Invoices are sometimes sent with the goods, but they are more usually mailed separately. If the buyer is not a regular customer, they may be expected to settle the account immediately. Regular customers will be given credit, with invoices being charged to their accounts. Payment will then be made later on the basis of a statement of account sent by the supplier monthly or at other periodic intervals.

An example of an invoice is shown in Figure 12.1.

JOHN G GARTSIDE & CO LTD

Albion Works, Thomas Street
Manchester M60 2QA
Telephone 0161-980-2132
INVOICE

Johnson Tools & Co Ltd
112 Kingsway
Liverpool
L20 6HJ

Your order no: AW 25

Date: 18 August 201-
Invoice no: B 832

Quantity	Item(s)	Unit Price	Total
			£
10	Polyester shirts, small	25.00	250.00
21	Polyester shirts, medium	26.00	546.00
12	Polyester shirts, large	27.25	327.00
			1,123.00
	VAT (@ 17.5%)		196.53
	One case (returnable)		23.25
			1,342.78
	Terms 21/2% one month		
E & OE		Registered in England No 523807	

FIGURE 12.1 Invoice

VAT: Value Added Tax. A tax on goods and services, payable to HM Customs and Excise.

E & OE: Errors and omissions excepted. This statement reserves the supplier's right to correct any errors which the document may contain.

Pro forma invoices

'Pro forma' means 'for form's sake'. A pro forma invoice is used:

- to cover goods sent 'on approval' or 'on consignment'
- to serve as a formal quotation
- to serve as a request for payment in advance for goods ordered by an unknown customer or a doubtful payer
- where the value of goods exported is required for customs purposes.

Pro forma invoices are not entered in the books of account and are not charged to the accounts of the persons to whom they are sent.

Covering email with invoice

It is not normally necessary to send a covering letter with an invoice. Sometimes, however, the customer requires the invoice to be sent by email, in which case a short and simple note may be sent with it.

The invoice informs the buyer of the amount due for goods supplied on credit.

Non-regular customer

Dear Mr Timms

We attach our invoice number B 832 for the polyester shirts ordered on 13 August.

The goods are available from stock and will be sent to you as soon as we receive the amount shown on the attached invoice, namely £1,342.78.

I hope to hear from you soon.

Regular customer

Dear John

Our invoice number B 832 is attached covering the polyester shirts you ordered on 13 August.

These shirts have been packed ready for despatch and are being sent to you, carriage paid, by rail. They should reach you within a few days.

All the best

Mark Robinson

 TIP You can learn more about expressions like 'carriage paid' in Chapter 10.

Debit and credit notes

If the supplier has undercharged the buyer, a debit note may be sent for the amount of the undercharge. A debit note is in the nature of a supplementary invoice.

If the supplier has overcharged the buyer, then a credit note is issued. Credit notes are also issued to buyers when they return either goods (for example, where they are unsuitable) or packing materials on which there is a rebate. Credit notes are usually printed in red to distinguish them from invoices and debit notes. Examples of credit and debit notes are shown in Figures 12.2 and 12.3.

JOHN G GARTSIDE & CO LTD		
Albion Works, Thomas Street Manchester M60 2QA Telephone 0161-980-2132 DEBIT NOTE		
Johnson Tools & Co Ltd 112 Kingsway Liverpool L20 6HJ		Date: 22 August 201- Debit Note No. D.75

Date	Details	Price
		£
18.8.1-	21 Polyester Shirts, medium charged on invoice number B 832 @ £26.00 each Should be £26.70 each Difference	 14.70
Registered in England No 523807		

FIGURE 12.2 Debit note

A debit note is sent by the supplier to a buyer who has been undercharged in the original invoice.

JOHN G GARTSIDE & CO LTD		
Albion Works, Thomas Street Manchester M60 2QA Telephone 0161-980-2132 CREDIT NOTE		
Johnson Tools & Co Ltd 112 Kingsway Liverpool L20 6HJ		Date: 25 August 201- Credit Note No. C.521

Date	Details	Price
18.8.1-	One case returned charged to you on invoice number B 832	£ 23.25
Registered in England No 523807		

FIGURE 12.3 Credit note

A credit note is sent by the supplier to a buyer who has been overcharged in the original invoice, or to acknowledge and allow credit for goods returned by the buyer. It is usually printed in red.

Supplier sends debit note

Dear Mr Lim

I am sorry that an error was made on our invoice number B 832 of 18 August.

The correct charge for polyester shirts, medium, is £26.70 and not £26.00 as stated. We are therefore enclosing a debit note for the amount undercharged, namely £14.70.

We are sorry this error was not noticed before the invoice was sent.

Yours faithfully

Buyer requests credit note

When notifying of an overcharge, some customers send a debit note to the supplier as a claim for the amount overcharged. If the supplier agrees to the claim, he will then issue a credit note to the customer.

Returned packing case

Dear Sirs

We have today returned to you by rail one empty packing case, charged on your invoice number B 832 of 18 August at £23.25.

A debit note for this amount is enclosed and we shall be glad to receive your credit note in return.

Yours faithfully

Incorrect trade discount

Dear Sir

Your invoice number 2370 dated 10 September allows a trade discount of only 33.3% instead of 40% as agreed in your letter dated 5 August in view of the unusually large order.

When calculated on the invoice gross total of £1,500, the difference in discount is exactly £100. Can you please adjust your charge and then we will pass the invoice for immediate payment.

Yours faithfully

Statements of account

A statement (see Figure 12.4) is a summary of the transactions between buyer and supplier during a specified period, usually one month. It starts with the balance

---

JOHN G GARTSIDE & CO LTD

Albion Works, Thomas Street
Manchester M60 2QA
Telephone 0161-980-2132
STATEMENT

Johnson Tools & Co Ltd
112 Kingsway
Liverpool
L20 6HJ

Date: 31 August 201-

Date	Details	Debit	Credit	Balance
		£	£	£
18.8.1-	Account rendered			115.53
18.8.1-	Invoice B 832	1,342.78		1,458.31
20.8.1-	Cheque received		500.00	958.31
22.8.1-	Debit Note D 75	35.70		994.01
25.8.1-	Credit Note C 52		23.25	970.76

E & OE

Registered in England No 523807

FIGURE 12.4 Statement

A statement is a demand for payment sent at regular periods by the supplier to buyers. It summarises all transactions over the period it covers and enables the buyer to check against the particulars given. Any errors discovered and agreed will be adjusted either by debit or credit note. Do check in your own country as to the minimum requirements for what to do if goods are faulty, and for returns and exchanges.

owing at the beginning of the period, if any. Amounts of invoices and debit notes issued are then listed, in chronological order, and amounts of any credit notes issued and payments made by the buyer are deducted. The closing balance shows the amount owing at the date of the statement.

Statements, like invoices, are generally sent without a covering letter. If a covering letter is sent, it need only be very short and formal.

Supplier reports underpaid statement

Supplier's letter

Dear Sirs

We are enclosing our September statement totalling £820.57.

The opening balance brought forward is the amount left uncovered by the cheque received from you against our August statement, which totalled £560.27. The cheque received from you, however, was drawn for £500.27 only, leaving the unpaid balance of £60 brought forward.

We look forward to receiving early settlement of the total amount now due.

Yours faithfully

Buyer's reply

Dear Sirs

We have received your letter of 15 October enclosing September's statement.

We apologise for the underpayment of £60 on your August statement. This was due to a misreading of the amount due. The final figure was not very clearly printed and we mistakenly read it as £500.27 instead of £560.27.

Our cheque for £820.57, the total amount on the September statement, is enclosed.

Yours faithfully

 TIP Notice the two commas in the final sentence. You can find out more about correct punctuation in Chapter 2.

Buyer reports errors in statement

Buyer's notification

Dear Sirs

On checking your statement for July we notice the following errors:

1. The sum of £14.10 for the return of empty packing cases, covered by your credit note number 621 dated 5 July, has not been entered.
2. Invoice Number W825 for £127.32 has been debited twice – once on 11 July and again on 21 July.

Therefore, we have deducted the sum of £141.42 from the balance shown on your statement. Our cheque for £354.50 is attached in full settlement.

Yours faithfully

Supplier's acknowledgement

Dear Sirs

Thank you for your letter of 10 August enclosing your cheque for £354.50 in full settlement of the amount due against our July statement.

Our apologies for the inconvenience.

Yours faithfully

Varying the terms of payment

When a customer is required to pay for goods when, or before, they are delivered, he is said to pay 'on invoice'. Customers known to be creditworthy may be granted 'open account' terms, under which invoices are charged to their accounts. Settlement is then made on the basis of statements of account sent by the supplier.

When a customer finds it necessary to ask for time to pay, the reasons given must be strong enough to convince the supplier that the difficulties are purely temporary and that payment will be made later.

Customer requests time to pay (granted)

Customer's request

Acknowledge supplier's letter

Explain why payment has not been made

Request deferred payment and give assurance

Apologise

Dear Sirs

We have received your letter of 6 August reminding us that payment of the amount owing on your June statement is overdue.

We were under the impression that payment was not due until the end of August when we would have had no difficulty in settling your account. However, it seems that we misunderstood your terms of payment.

In the circumstances, we are hoping you will allow us to defer payment for a further three weeks. Our present difficulty is purely temporary. Payments are due to us before the end of the month from a number of our regular customers who are notably prompt payers.

We are sorry to make this request and hope to receive your positive reply soon.

Yours faithfully

 TIP Notice the considerate tone in this message. You can learn more about using the right tone in Chapter 3.

Supplier's reply

Respond to request for deferred payment

Explain reason for granting request

Reminder about future terms

Dear Mr Jensen

We have considered your letter of 8 August, and have decided to allow you to defer payment of your account to the end of August.

We are granting this request because you have normally settled your accounts promptly in the past. We hope that in future dealings you will be able to keep to our terms of payment, which are:

2.5% discount for payment within 10 days

Net cash for payment within one month

We look forward to continuing to work with you.

Yours sincerely

Customer requests time to pay (not granted)

Customer's request

Dear Mr Wilson

Thank you for your letter dated 23 July asking for immediate payment of the £1,987 due on your invoice number AV54.

When we wrote promising to pay you in full by 16 July, we fully expected to be able to do so. However, we were called upon to meet an unforeseen and unusually heavy demand earlier this month.

We are enclosing a cheque for £1,000 on account, and hope you will allow us a further few weeks in which to settle the balance. We fully expect to be able to settle your account in full by the end of August.

I hope to hear from you soon.

Yours sincerely

CAUTION Take care when using the word 'However'. Many people make mistakes with punctuation. It is correctly used in the second paragraph of this message. Learn more about punctuation in Chapter 2.

Supplier's reply

In refusing requests of this kind it is better for suppliers to stress the benefits the customer is likely to gain from making payments promptly rather than to stress their own difficulties in seeking prompt payment. The customer is, after all, more interested in problems closer to home.

Dear Mrs Billingham

Thank you —— Thank you for your letter dated 25 July sending us a cheque for £1,000 on account and asking for an extension of time in which to pay the balance.

Tactfully state that —— As your account is now more than two months overdue, we do payment is insufficient not feel it is reasonable to expect us to wait another month for the and the delay quite balance, particularly as we invoiced the goods at a specially low unreasonable price, which was mentioned to you at the time.

Express sympathy but —— We are sorry to hear of your difficulties, but need hardly remind explain why prompt you that it is in our customers' long-term interests to pay their payment is necessary accounts promptly so as to qualify for discounts. This helps to build a reputation for financial reliability.

The request for —— In the circumstances, we hope you will be able to clear your immediate payment is account without further delay. We look forward to receiving your worded appropriately cheque soon.

Yours sincerely

 CAUTION Take care not to use a tone that would offend. For more guidance on using the right tone, see Chapter 3.

Supplier questions partial payment

When making payment on a statement, the debtor should always state whether the payment is 'on account' or 'in full settlement', otherwise it may give rise to letters such as this:

Dear

Thank you for your letter dated 10 October enclosing your cheque for £58.67. Our official receipt is enclosed as requested.

As you do not say that the cheque is on account, we are wondering if this amount was intended to be £88.67 – this is the balance on your account as shown in our September statement.

We look forward to receiving the uncleared balance of £30 within the next few days.

Yours sincerely

Supplier rejects discount deduction

Dear

Thank you for your letter dated 15 October stating that your cheque for £282.50 is in full settlement of your May statement.

Unfortunately, we cannot accept this payment as full payment of the £300 due on our statement. The terms of payment allow the 2.5% cash discount only on accounts paid within 10 days of the statement, but your recent payment is more than a month overdue.

The balance still owing is £17.50 and to save you the trouble of making a separate payment we will include this amount in your next statement.

Yours sincerely

Methods of payment

Various methods of payment may be used in settling accounts. The type of payment to be used is a matter for agreement between the parties concerned. This book is not intended to be a resource for details of this kind, but it was felt important to provide some details here of various methods. It is, of course, advisable to check in your own country about all the methods available.

1 Payments through banks

Online banking

More and more people are paying bills using digital banking services on the Internet. After setting up your unique username and passwords, you can pay bills, transfer money, do virtually anything with your accounts at the click of a mouse.

Cheques

A bank cheque is always payable on demand. It is by far the most common form of payment used to settle credit transactions in the home trade of countries where the bank cheque system has been developed. It may also be used to pay debts abroad. A receipt is the best, but not the only, evidence of payment. Cheques that have been paid by a banker and later returned to customers may be produced as receipts. When payment is made by cheque, a separate receipt is therefore unnecessary but the payer may legally demand a receipt if required.

Bank giro credit

Bank giro credits (BGCs) are used by customers to pay cash or cheques into a bank account. They are commonly found in the form of tear-off strips at the bottom of utility, telephone and other regular bills. A bank giro credit is basically a paper slip addressed to a bank branch instructing it to credit a specified sum of money to a named account at that branch. It bears the money mark logo next to the words 'bank giro credit'. A bank giro credit is not a payment instrument, ie it cannot be used on its own to make a payment, and must be accompanied by cash and/or cheque – so the use of bank giro credits tends to follow any trends in the use of cheques and cash as a method of payment.

You can use a bank giro credit to pay a bill in two ways:

- by post with a cheque
- by paying the amount into a branch of the paying customer's own bank with cash or cheque.

Bank giro credits can also be found in the back of cheque books, and are used by customers to pay cash or cheques into their own bank accounts.

FPS or BACS

Payments within the UK are can be sent via the Faster Payments Service (FPS) or by BACS, depending on how soon the funds are needed. If sending an FPS or BACS payment, you must have the receiver's account details. If sending money abroad, you can request this from your bank providing you have the International Bank Account Number (IBAN) and SWIFT BIC (Bank Identifier Code).

Electronic Funds Transfer (EFT)

Electronic Funds Transfer (EFT) is a system of transferring money from one bank account directly to another without any paper money changing hands. One of the most widely used EFT programs is Direct Deposit, in which payroll is deposited straight into an employee's bank account, although EFT refers to any transfer of funds initiated through an electronic terminal, including credit card, ATM, Fedwire and point-of-sale (POS) transactions. It is used for both credit transfers, such as payroll payments, and for debit transfers, such as mortgage payments.

EFT transactions are processed by banks, and a fee is usually charged. Funds are transferred electronically from one bank account to the billing company's bank, usually less than a day after the scheduled payment date. The growing popularity of EFT for online bill payment is paving the way for a paperless universe where cheques, stamps, envelopes and paper bills are obsolete. The benefits of EFT include reduced administrative costs, increased efficiency, simplified bookkeeping and greater security.

Banker's drafts

A banker's draft is a document bought from a bank, for which the bank will usually charge a fee. It orders the branch bank, or the agent on whom it is drawn, to pay the stated sum of money on demand to the person named in the draft (the payee). In foreign transactions the payee receives payment in the local currency at the current rate of exchange. Banker's drafts are convenient for paying large sums of money in circumstances where a creditor would hesitate to take a cheque in payment. Like cheques, they may be crossed for added safety.

2 Cash (coins and notes)

If paying or receiving cash for goods, you should ensure that a proper receipt is provided for inventory and accounting purposes. The vendor should also keep a record of these transactions.

3 PayPal (and other peer-to-peer payment websites)

PayPal is an e-commerce business that enables you to pay, send money and accept payments without revealing your financial information. In short, PayPal acts like a digital wallet where you can securely store all your payment options, such as your bank account and credit card details. When you want to make a payment, you don't have to pull out your credit card or type your billing info every time. PayPal is a safer, faster, more secure way to pay online – your financial details are never shared, so your bank, credit and debit card details are safeguarded. Anyone can use PayPal to pay for items with any of the thousands of retailers that accept PayPal.

4 Western Union (and other wire transfer services)

Western Union Money Transfer enables quick and easy money transfer to most locations worldwide. In order to send funds, a sender goes to a Western Union office and presents funds (plus fees). A sender provides his or her name and address, the recipient's name, and a designated payment destination. Western Union then gives the sender a 10-digit Money Transfer Control Number (MTCN) that must be transmitted separately by the sender to the recipient. The recipient then proceeds to a Western Union agent office in the designated payment location, presents the 10-digit MTCN, and a photo ID. Money is then paid out to the recipient. If a recipient lacks identification documents, the sender and receiver can choose to set up a prearranged password. Funds are paid out in cash, although if payment exceeds a local maximum or cash on hand, a cheque is issued. Alternatively, a sender may forward funds online to a recipient by using Western Union's online site, **www.westernunion.com**.

5 Payments through the Post Office

Postal orders

British postal orders are similar to cheques for those without bank accounts, or perhaps you want to order online and are nervous about giving your personal financial details. Postal orders look just like cheques with the payment amount printed in words and numbers. They are available in any value up to UK £250, and you may have the payee name printed on the postal order, making them even more secure. With each postal order, you'll receive a separate receipt containing all the information you need should you wish to make an enquiry. Many countries overseas will cash British postal orders, and a full list is available on the Internet.

MoneyGram®

MoneyGram® are located in over 233,000 agent locations in 190 countries and territories worldwide. These allow you to securely transfer up to UK £5,000, and the money will typically arrive within 10 minutes. Simply take some personal identification with you to the post office, complete a simple form and hand it in together with your cash and transfer fee. Contact the recipient and give them your reference number. Within 10 minutes the recipient will be able to visit their local post office with some identification, quote the reference number, and receive the money.

Giro transfers

Giro transfer is a payment transfer from one bank account to another bank account, instigated by the payer, not the payee. Most countries have giro transfer systems, sometimes referred to under different names. These electronic payments are considered to be faster, cheaper and safer due to the reduced risk of fraud.

Supplier asks customer to select terms of payment

Dear

Thank you for your letter of 3 April, but you do not say whether you wish this transaction to be for cash or on credit.

When we wrote to you on 20 March we explained our willingness to offer easy credit terms to customers who do not wish to pay cash, and also that we allow generous discounts to cash customers.

We may not have made it clear that when placing orders customers should state whether cash or credit terms are required.

Please let me know which you prefer so that we can set up your account accordingly.

Yours sincerely

Form letter enclosing payment (and acknowledgement)

Every business has a good deal of purely routine correspondence, which can often be dealt with by template letters. Letters enclosing or acknowledging payments fall into this category. They often take a standard form suitable for all occasions and are therefore known as 'form letters' or templates. In such cases a supply of pre-printed letters is often prepared with blank spaces for the variable information to be completed (reference numbers, names and addresses, dates, sums of money, etc).

The personal touch that personalised letters provide is lost with form letters. However, many companies now use mail merge facilities to produce personalised form letters that look like originals.

Sender's form letter

Dear Sirs

We have pleasure in enclosing our cheque (bill/draft/etc) for £___ in full settlement (part settlement) of your statement (invoice) dated.

Please send us your official receipt.

Yours faithfully

Form letter acknowledging payment

> Dear
>
> Thank you for your letter of ___ enclosing cheque (bill/draft/etc) for £___ in full settlement (part payment) of our statement of account (invoice) dated ___.
>
> We enclose our official receipt.
>
> Yours sincerely

Letter informing supplier of payment by credit transfer

> Dear Sirs
>
> YOUR INVOICE NUMBER 1524
>
> We have today made a credit transfer to your account at the Barminster Bank, Church Street, Dover, in payment of the amount due for the goods supplied on 2 May.
>
> Yours faithfully

Letter informing supplier of payment by banker's draft

> Dear Sirs
>
> Our banker's draft is enclosed, drawn on the Midminster Bank, Benghazi, for £2,672.72 and crossed 'Account Payee only'.
>
> The draft is sent in full settlement of your account dated 31 May.
>
> Please acknowledge receipt.
>
> Yours faithfully

Useful expressions

Payments due

Openings

■ Enclosed is our statement for the quarter ended ___.

■ We enclose our statement to 31 ___ showing a balance of £___.

■ We are sorry it was necessary to return our invoice number ___ for correction.

Closes

■ Please let us have your credit note for the amount of this overcharge.

■ Please make the necessary adjustment and we will settle the account immediately.

■ We apologise again for this error and enclose our credit note for the overcharge.

Payments made

Openings

■ We enclose our cheque for £___ in payment for goods supplied on ___.

■ We enclose our cheque for £___ in settlement of your invoice number ___.

■ Many thanks for your cheque for £___.

■ We thank you for your cheque for £___ in part payment of your account.

Closes

■ We hope to receive the amount due by the end of this month.

■ Please send us your cheque immediately.

■ As the amount owing is considerably overdue, we must ask you to send us your cheque by return.

13
Letters requesting payment

Tone

When a customer does not pay an account promptly, it is always annoying to the supplier. It's important, however, to write in a very courteous manner rather than allow any suggestion of annoyance to creep into the correspondence. It may be better not to write at all and instead telephone tactfully to find out why the payment has not been made. In difficult cases it may even be good policy to accept a part payment rather than resort to legal action, which would be both expensive and time-consuming.

 CAUTION There may be several good reasons why a customer fails to pay on time, some of them deserving sympathy. There is, however, always the customer who is only too ready to invent excuses and who needs to be monitored. Each case must be treated individually.

The style and tone of any letters should depend on such factors as the age of the debt, whether late payment is habitual, and how important the customer is. However, no letter must ever be less than courteous.

 TIP For more on tone see Chapter 3.

Late payments

When a debtor needs to write explaining difficulties in paying an account by the due date, with a request to defer payment, the following plan is useful:

1 Refer to the account that cannot be paid immediately.

2 Apologise for inability to pay and give reasons.

3 Suggest an extension of period for payment.

4 Hope that the suggestion will be accepted.

Customer explains inability to pay

This letter is from a regular and reliable customer who makes a reasonable request. If the supplier refuses this, it could run the risk of driving the customer away. The customer might pay the outstanding amount, but could then start buying from a competitor. In the process the supplier could lose many valuable future orders.

Dear Sirs

We have received your invoice number 527 dated 20 July for £10,516, which is due for payment at the end of this month.

Unfortunately, a fire broke out in our Despatch Department last week and destroyed a large part of a valuable consignment due for delivery to a cash customer. Our claim is now with the insurance company, but it is unlikely to be met for another three or four weeks. Until then we are faced with a difficult financial problem.

I am hoping you will allow us to defer payment of your invoice until the end of September.

As you are aware, our accounts with you have always been settled promptly, and we hope you will be able to agree to this extension.

If you have any questions please give me a call.

Yours faithfully

Customer explains late payment

Dear

I am sorry for the late payment of your invoice number W563, and am now enclosing our cheque for £1,182.57 in full settlement.

The delay was due to my absence from the office through illness, and I did not leave instructions for your account to be paid. I only discovered the oversight when I returned to the office yesterday.

My apologies once again for this delay.

Yours sincerely

Collection letters

The preliminary steps in debt collection are:

1. A first end-of-month statement of account.
2. A second end-of-month statement of account with added comment.
3. A first collection letter worded formally.
4. Second and third collection letters.
5. A final letter notifying that legal action will be taken unless the amount is paid within a stipulated period of time.

A customer whose account is only slightly overdue would understandably be offended to receive a personal letter about this. This is why the first two reminders usually take the form of end-of-the-month statements of account. The second of these statements is usually marked with such comments as 'Second application', 'Account overdue – please pay' or 'Immediate attention requested'.

If payment has still not been received after sending two copy statements, the supplier must send a letter requesting payment, which are called 'collection letters'. These collection letters aim to:

- persuade the customer to settle the account
- retain custom and goodwill.

CAUTION Without using appropriate wording and tone, it would be easy to cause offence, so collection letters must be written with tact and restraint. Remember also that the supplier could actually be at fault, for example if a payment received has not been recorded, or if goods sent or services provided are not satisfactory.

First applications for payment

A printed collection letter

A first collection letter may be printed as a 'form letter' as in this example where the individual details are merged appropriately, personalising the letter.

> Dear Sir/Madam
>
> ACCOUNT NUMBER ___
>
> According to our records you have not settled this account, despite several reminders.
>
> The enclosed statement shows the amount owing to be £___.
>
> We hope to receive an early settlement of this account.
>
> Yours faithfully

Personalised collection letters

There may be circumstances when an individual letter rather than a form letter is more appropriate. It should then be addressed to a named senior official and marked 'Confidential'.

To a regular payee

Dear Mr Farrelly

ACCOUNT NUMBER 6251

As you are usually very prompt in settling your accounts, we wonder whether there is any special reason why we have not received payment of this account, which is already two months overdue.

We are attaching another statement of account dated 31 May showing a balance owing of £1,505.67.

We hope to receive prompt payment.

Yours sincerely

To a new customer

Dear Mrs Yap

ACCOUNT NUMBER 5768

Unfortunately, we have not received payment of the balance of £560.55 due on our statement for December. A copy is enclosed.

We must remind you that unusually low prices were quoted to you on the under-standing that you would make early settlement.

I'm sure this is an oversight, and hope to receive your payment within the next few days.

Yours faithfully

To a customer who has sent a part payment

Dear

Thank you for your letter of 8 March enclosing a cheque for £500 in part payment of the balance due on our February statement.

Your payment leaves an unpaid balance of £825.62. As our policy is to work on small profit margins, we are unable to grant long-term credit facilities.

We are sure you will understand if we ask for immediate payment of this balance.

Yours sincerely

Reminder to customer who has already paid

The need for a cautious approach is always necessary since the customer may not be at fault, as where the payment has gone missing or where the supplier has received it but has not recorded it.

Request for payment

Dear

ACCOUNT NUMBER S542

According to our records, our account for cutlery supplied to you on 21 October has not been paid.

We enclose a detailed statement showing the amount owing to be £2,310.62, and hope to receive prompt payment.

Yours faithfully

Customer's reply

Dear

YOUR ACCOUNT NUMBER S542

I was surprised to receive your letter of 8 December stating that you had not received payment of this account.

Our cheque (RBS, number 065821) for £2,310.62 was mailed to you on 3 November. As you have obviously not received this, I have instructed our bank to cancel this cheque. A replacement cheque for the same amount is enclosed.

Yours sincerely

Second application letters

If a reply to the first application is not received, a second letter should be sent after about 10 days. This should be firmer in tone but still polite. Co-operation and good relations will not be achieved by annoying the customer.

Such letters should be addressed to a senior official and planned as follows:

1. Refer to previous letter.
2. Assume that something unusual accounts for the delay in payment.
3. Suggest tactfully that an explanation would be welcome.
4. Ask for payment to be sent.

Specimen second application letters

Second letter, following letter on page 193

Dear Mr Farrelly

ACCOUNT NUMBER 6251

We have not received a reply to our letter of 5 July requesting settlement of this account. The amount still owing is £1,505.67.

No doubt there is some special reason for the delay in payment, and we should welcome an explanation together with your immediate payment.

Yours sincerely

Second letter, following letter on page 193

Dear Mrs Yap

We wrote to you on 18 February to remind you that a total of £560.55 was outstanding and due for payment by 31 January.

This account is now more than two months overdue, so we must ask you either to send us your payment within the next few days or at least to offer an explanation for the delay in payment.

I look forward to your prompt reply.

Yours faithfully

Second letter, following letter on page 194

Dear

We have not heard from you since we wrote on 10 March about your unpaid balance of £825.62. In view of your past good record, we have not previously pressed for a settlement.

To regular customers like you, our terms of payment are 3% one month. We hope you will not withhold payment any longer, otherwise it will be necessary for us to revise these terms.

We look forward to receiving your cheque for the outstanding amount within the next few days.

Yours sincerely

Third application letters

If payment is still not made and if no explanation has been received, a third letter becomes necessary. Such a letter should show that steps will be taken to enforce payment if necessary, such steps depending on individual circumstances. Third letters should follow this plan:

1 Review earlier efforts to collect payment.

2 Give a final opportunity to pay by stating a reasonable deadline date.

3 State that you wish to be fair and reasonable.

4 State action to be taken if this third request is ignored.

Specimen third application letters

Third letter, following letter on page 195

Dear Mr Farrelly

ACCOUNT NUMBER 6251

We do not appear to have received replies to our two previous requests of 5 and 16 July for payment of £1,505.67 still owing on this account.

We have no wish to be unreasonable, but at this stage we feel we must insist on immediate payment. If we do not receive payment by 7 August, we will have no choice but to take other steps to recover this amount.

We sincerely hope this will not become necessary.

Yours sincerely

Third letter, following letter on page 196

Dear Mrs Yap

Even in this third letter restraint is shown in the wording rather than directly attacking the customer — I am sorry we have not heard from you in reply to our two letters of 18 February and 2 March regarding the £560.55 due on our December statement. We had hoped that you would at least explain why the account continues to remain unpaid.

Terms like 'every consideration' and 'no choice' somewhat soften the blow — I am sure you will agree that we have shown every consideration in the circumstances. Unfortunately, we now have no further choice but to take other steps to recover the amount due.

This paragraph gives the customer a final chance to clear the account — We are most anxious to avoid doing anything through which your credit and reputation might suffer. Therefore, even at this late stage we are prepared to give you a further opportunity to put matters right.

A specific timeframe is given — We are willing to give you until the end of this month to clear your account.

Yours faithfully

Third letter, following letter on page 196

> Dear
>
> We are disappointed not to have heard from you since our two letters of 10 and 23 March reminding you of the balance of £825.62 still owing on our February statement.
>
> This is all the more disappointing because of our past good relationship over many years.
>
> In the circumstances, unless we hear from you within 10 days we shall have to consider seriously the further steps we could take to obtain payment.
>
> Yours sincerely

Final collection letters

If all three applications are ignored, it is reasonable to assume that the customer either cannot, or will not, settle the account. A brief notification of the action that is to be taken must then be sent as a final warning.

Specimen final collection letters

Final letter, following letter on page 197

> Dear Mr Farrelly
>
> We are surprised that we have received no reply to the further letter we sent to you on 28 July regarding the long overdue payment of £1,505.67.
>
> Our relations in the past have always been good. Even so, we cannot allow this amount to remain unpaid indefinitely. Unless the amount due is paid or a satisfactory explanation received by the end of this month, we shall be reluctantly compelled to put this matter in the hands of our solicitors.
>
> Yours sincerely

Final letter, following letter on page 197

Dear Mrs Yap

We are disappointed not to have received any response from you in answer to our letter of 16 March concerning non-payment of the balance of £560.55 outstanding on our December statement.

We are now making a final request for payment in the hope that it will not be necessary to hand the matter over to an agent for collection.

We have decided to defer this step for 7 days to give you the opportunity either to pay or at least to send us an explanation.

Yours faithfully

Final letter, following letter on page 198

Dear

We are unable to understand why we have received no reply to our letter of 7 April, our third attempt to secure payment of the balance of £825.62 still owing on your account with us.

We feel that we have shown reasonable patience and treated you with every consideration. However, we must now take steps to recover payment at law, and the matter will be placed in the hands of our solicitors.

Yours sincerely

Useful expressions

First applications

- We notice that your account, which was due for payment on ___, is still outstanding.
- It seems that our invoice number ___ for £___ remains unpaid.
- We have not yet received the balance of our ___ statement amounting to £___.

- As our statement may have gone astray, a copy is attached. Please pass this for immediate payment.
- We hope to receive your cheque by return.
- We look forward to receiving your payment soon.

Second applications

- We have not received a reply to our letter dated ___ for settlement of £___ .
- We are disappointed not to have received a reply to our letter dated ___ requesting settlement of our statement.
- We hope you will deal with this without further delay.
- We must ask you to settle this account immediately.
- We hope you will make arrangements to pay the amount outstanding immediately.

Third applications

- We wrote to you on ___ and again on ___ regarding the amount owing on our invoice number ___ .
- We have had no reply to our previous requests for payment of our ___ statement.
- We are disappointed that we have had no replies to our two previous applications for payment of your outstanding account.
- Unless we receive your cheque in full settlement by ___ we shall have no option but to instruct our solicitors to recover the amount due.
- Unless we receive your full payment by the end of this month, we shall be compelled to take further steps to enforce payment.
- We still hope you will settle this account without further delay, otherwise we shall have to take legal action.

Checklist

■ Use a tone that is firm but understanding.

■ Take care not to cause offence.

■ Mention when the payment was originally due.

■ State the amount owed.

■ State the penalties if any.

■ Mention the grace period.

■ Give a new deadline.

■ Indicate the consequences.

■ Be considerate and at the same time firm.

■ Refer to Chapter 3 for more guidance on tone.

14 Credit and status enquiries

Reasons for credit

The main reason for buying on credit is for convenience. Basically it allows us to 'buy now, pay later'.

1 Credit enables a retailer to hold stocks and to pay for them out of the proceeds of later sales. This increases the working capital and helps to finance the business.

2 Credit enables the buying public to enjoy the use of goods before they have saved the money needed to buy them.

3 Credit avoids the inconvenience of separate payments each time a purchase is made.

The main reason for selling on credit is to increase profits. Credit sales not only attract new customers but also keep old customers, since people who run accounts tend to shop at the place where the account is kept, whereas cash customers are free to shop anywhere.

Disadvantages of credit

There are a number of disadvantages in dealing on credit both for the supplier and for the customer:

1 It increases the cost of doing business since it involves extra work in keeping records and collecting payments.

2 It exposes the supplier to the risk of bad debts.

3 The buyer pays more for the goods since the supplier must raise prices to cover the higher costs.

Requests for credit

A buyer who makes regular purchases from the same supplier will usually wish to avoid the inconvenience of paying for each transaction separately, and will ask for 'open account' terms under which purchases will be paid for monthly or quarterly or at some other agreed period. In other words, the goods are to be supplied on credit.

Customer requests open account terms (credit terms with periodic settlement)

Request

Dear

We have been very satisfied with your handling of our orders over the past two years. As our business is growing, we expect to place even larger orders with you in the future.

We hope you will agree to allow us open account facilities with, say, quarterly settlements. This arrangement would save us the inconvenience of making separate payments on invoice.

If necessary, we will be happy to supply banker's and trade references.

We hope to hear from you soon.

Yours sincerely

Reply accepting request

Dear

Thank you for your letter dated 18 November requesting open account terms.

We are quite happy to make this change, based on a 90-day settlement period. In your case it will not be necessary to supply references.

We are pleased that you have been satisfied with our past service and that expansion of your business is likely to lead to increased orders.

We can assure you of our continued efforts to give you the same high standard of service as in the past.

Yours sincerely

Customer requests extension of credit

Cash flow problem

Dear

We are sorry that you had to remind us that we have not settled your account due for payment on 30 October.

We had intended to settle this account earlier. However, unfortunately our own customers have not been meeting their obligations as promptly as usual. This has had a big impact on our cash flow.

Investment income due in less than a month's time will enable us to clear your account by the end of next month. We hope you can accept the enclosed cheque for £200 as a payment on account. The balance will be cleared as soon as possible.

Yours sincerely

Lending restrictions and bad trade

Dear

STATEMENT OF ACCOUNT FOR AUGUST 201-

We have just received your letter of 8 October requesting settlement of our outstanding balance of £1,686.00.

We are sorry not to have been able to clear this balance as quickly as usual. However, the present depressed state of business and the current restrictions on bank lending have created difficulties for us. These difficulties are purely temporary as payments from customers will come in early in the New Year on a number of recently completed contracts.

Our resources are quite sufficient to meet all our obligations, but I'm sure you will appreciate we have no wish to realise on our assets at the moment. We hope you will therefore grant us a three-month extension of credit, when we will be able to settle your account in full.

Yours sincerely

Customer requests credit extension due to bankruptcy

Letter to supplier

Dear

Introduction gives background details — We have received your statement for the quarter ended 30 September showing a balance due £785.72.

History of prompt payment is explained with details of current situation — Until now we have always settled our accounts with you promptly. We could have done so at this time too if it wasn't for the bankruptcy of an important customer whose affairs are not likely to be settled for some time.

Tactful request to defer payment — We hope you can allow us to defer payment of your present account to the end of next month. This would enable us to meet a temporarily difficult situation forced upon us by events that could not be foreseen.

Final assurance of early settlement — During the next few weeks we will be receiving payments under a number of large contracts. We shall then have no difficulty in settling with you in full in due course.

If you wish to discuss this please give me a call on 2468742.

Yours sincerely

Request granted

Refer to customer's letter and request — Dear

Thank you for your letter of 10 October requesting an extension of time to settle of our 30 September statement.

State reason for agreeing to extension — In view of the promptness with which you have always settled with us in the past, we are willing to grant this extension in these special circumstances.

Give a final date for full settlement — Please let us have your cheque in full settlement by 30 November.

Yours sincerely

Request refused

Refer to customer's letter and request — Dear

Thank you for your letter of 10 October. I am sorry to learn about the difficult situation in which the bankruptcy of an important customer has placed you.

Tactful wording is necessary when a request is refused — We fully understand your wish for an extension of time and would like to be able to help you. Unfortunately, this is impossible because of commitments that we must meet by the end of this month.

Explain regret at requesting immediate payment — Your request is not at all unreasonable and if it had been possible we would have been pleased to grant it. In the circumstances, however, we must ask you to settle with us on the terms of payment originally agreed.

Yours sincerely

Business references

When goods are sold for cash, there is no need for the supplier to enquire into the financial standing of the buyer. Where they are sold on credit, however, the ability to pay will be important.

For credit to be allowed, the supplier will want to know details about the buyer's reputation, the extent of their business, and in particular whether accounts are paid promptly. It is on this information that the supplier will decide whether to allow credit and, if so, how much.

This information can be obtained from:

- trade references supplied by the customer
- the customer's banker
- various trade associations
- credit enquiry agencies.

When a customer places an order with a new supplier, it is usual to supply trade references, that is the names of persons or firms to whom the supplier may refer for information. Alternatively or additionally, the customer may give the name and address of the banker.

 CAUTION As these references are provided by customers, they must be accepted with caution since naturally only those who are likely to report favourably will be named as referees. Even a bank reference can be misleading. A customer may have a satisfactory banking account and yet have dubious business dealings.

Supplier requests references

When a new customer places an order but fails to provide references, the supplier will naturally want some evidence of the customer's creditworthiness, especially for a large order. The supplier's letter asking for references must avoid any suggestion that the customer is not to be trusted.

Dear

We were pleased to receive your first order with us dated 19 May.

When opening new accounts, it is our usual practice to ask customers for trade references. Please send us the names and addresses of two other suppliers with whom you have regular dealings.

We hope to hear from you soon. Meanwhile, your order has been put in hand for despatch immediately we hear from you.

Yours sincerely

Supplier asks for completion of credit application form

Letter from supplier

> Dear
>
> Thank you for your order number 526 of 15 June for polyester bedspreads and pillow cases.
>
> As you are a new customer, we should like you to take advantage of our usual credit terms, so our credit application form is enclosed. Please complete this and return to me.
>
> We should be able to deliver your present order in about two weeks, and look forward to receiving your further orders.
>
> We hope that this first transaction will mark the beginning of a great business relationship between us.
>
> Yours sincerely

 TIP Notice how all these letters are written in accordance with the four-point plan. See Chapter 4 for more on how to structure messages logically.

Customer returns completed credit application form

> Dear
>
> Thank you for your letter of 18 June.
>
> We quite understand the need for references and have completed your credit application form giving the relevant information. This is enclosed.
>
> Please arrange for our first order to be delivered by the end of this month.
>
> We look forward to our future business dealings with you.
>
> Yours sincerely

Customer supplies trade references

Dear Sirs

Thank you for the catalogue and price list received earlier this month.

We have pleasure in sending you our first order, number ST6868, and hope you will allow us 25% discount on your usual monthly terms.

The equipment on this order is needed urgently by our customers. As we understand you have them in stock, we would like delivery by the end of next week. We hope this will leave enough time for you to take up references with these organisations with whom we have been working for many years:

1. Kisby & Co Ltd, 28–30 Lytham Square, Liverpool
2. Atlas Manufacturing Co Ltd, Century House, Bristol

We look forward to doing further business with you in the future.

Yours faithfully

Customer supplies a banker's reference

Dear Sirs

Our cheque for £2,513 is enclosed in full settlement of your invoice number 826.

My directors have good reason to believe that these products will be a popular selling line in this part of the country. As we expect to place regular orders with you, we hope you will agree to provide open account facilities on a quarterly basis.

For information about our credit standing please contact Barclays Bank Ltd, 25–27 The Arcade, Southampton.

We hope to hear from you soon.

Yours faithfully

Status enquiries

Letters taking up trade references are written in formal, polite terms. They usually conform to the following four-point plan:

- Give background information about the customer's situation.
- Request information about the prospective customer's standing and an opinion on the wisdom of granting credit within a stated limit.
- Give an assurance that the information will be treated confidentially.
- Enclose a stamped addressed envelope or an international postal reply coupon if the correspondent lives abroad.

Some large firms make their enquiries on a specially printed form containing the questions they would like answered. Use of such forms makes it easier for the companies approached, and helps to ensure prompt replies.

When the supplier receives the information requested, it is courteous to send a suitable letter of acknowledgement and thanks.

 CAUTION Always address letters taking up references to a senior official and mark them 'Confidential'.

Supplier takes up trade references

Example 1

> Dear Sirs
>
> Watson & Jones of Newcastle wish to open an account with us and have given your name as a reference.
>
> Please let us have your view on the firm's general standing and your opinion on whether they will be able to settle their accounts promptly with a credit up to £10,000.
>
> We will, of course, treat all information in strict confidence.
>
> We enclose a stamped, addressed envelope for your reply.
>
> Yours faithfully

Example 2

Dear Sirs

We have received a request from Shamlan & Shamlan & Co of Bahrain for supplies of our products on open account terms. They state that they have regularly traded with you over the past two years and have given your name as a reference.

Please let us know in confidence whether you have found this company to be reliable in their dealings with you and prompt in settling their accounts.

We understand their requirements with us may amount to approximately £5,000 a quarter. We should be glad to know if you feel they can meet such commitments.

We enclose an international postal reply coupon, and your reply will be treated in strict confidence.

Yours faithfully

Supplier requests his banker to take up bank reference

In view of the highly confidential relationship between bankers and their customers, a banker will not normally reply direct to private enquiries about a client's standing. This information is usually given willingly to fellow bankers. When taking up a bank reference, the supplier must do so through their own banker.

Dear Sir/Madam

The Colston Engineering Co Ltd in Mumbai has asked for us to grant them a standing credit of £20,000. However, as our knowledge of this company is limited to a few months' trading on the basis of cash-on-invoice, we would like some information about their financial standing before proceeding.

The only reference they give us is that of their bankers – the National Bank of Nigeria, Ibadan.

We hope you can let us have any information about this company that would help us in our decision.

Yours faithfully

Supplier refers to credit enquiry agency

A supplier who wants an independent reference about a customer's business standing may refer either to a trade association or to one of the numerous credit enquiry agencies. These agencies make it their business to supply information on the financial standing of both trading firms and professional and private individuals. They have a remarkable store of information, which is kept up to date from a variety of sources including their own local agents. If the information requested is not immediately available from their records, they will set up enquiries and can usually supply it within a few days.

Dear Sirs

We have received a first order worth £12,750 from A Griffiths & Co, Cardiff, who have requested open account terms.

We have no information about this company, but as there are prospects of further large orders we should like to meet this order and provide open account terms if it is safe to do so.

Please let us have a report on the reputation and financial standing of the company and whether it is advisable for us to grant credit for this first order. We would also appreciate advice on the maximum amount for which it would be safe to grant credit on a quarterly account.

Thanks for your help in advance.

Yours faithfully

Replies to status enquiries

Where a company's credit has been found to be satisfactory, the reply to the enquiry presents no problem. However, if the firm's credit is uncertain, the reply calls for the utmost care. It is usual to phrase such replies in a manner that leaves the enquirer to 'read between the lines', ie to gather for themselves the true meaning, rather than bluntly state disparaging facts.

Replies to letters taking up references should be marked 'Confidential' and follow this four-point plan:

■ Acknowledge the request and give background information.
■ State the facts and offer an honest expression of opinion.

- Hope that the information supplied will be useful.
- Tactfully remind them that the information is confidential and that you cannot accept any responsibility for it.

Trader's replies to credit information enquiry

Favourable reply to letter on page 211

Dear

Thank you for your letter of 25 May.

Watson & Jones of Newcastle is a small but well-known and highly respectable firm that has been established in this town for more than 25 years.

We have been doing business with them for over seven years on quarterly-account terms. Although they have not usually taken advantage of cash discounts, they have always paid their account promptly on the net dates. The credit we have allowed this company has at times been well over the £10,000 you mention.

We hope this information will be helpful and that it will be treated as confidential.

Yours sincerely

Discouraging reply to letter on page 212

Dear

Thank you for your letter of 25 May regarding Shamlan and Shamlan & Co of Bahrain.

This company has placed regular orders with us for several years. We believe the company is trustworthy and reliable, but we have to say that they have not always settled their accounts by the due date.

Their account with us is on quarterly settlement terms but we have never allowed it to reach the sum mentioned in your letter. I would advise that caution is necessary.

We are glad to be of help and ask you to treat this information as strictly confidential.

Yours sincerely

TIP Take pride in composing effective messages that are structured logically. See Chapter 4 for more details.

Banker's replies to credit information enquiry

Favourable reply to letter on page 212

Dear

We have received information from the National Bank of Nigeria regarding your letter dated 18 September.

The Colston Engineering Co Ltd in Mumbai is a private company founded 15 years ago and run as a family concern by three brothers. They enjoy a good reputation and punctually meet their commitments. It would seem to be safe to grant credit in the sum you mention.

This information is strictly confidential and is given without any responsibility on our part.

Yours sincerely

Unfavourable reply to letter on page 212

Dear

We have received information from the National Bank of Nigeria regarding your letter dated 18 September.

The Colston Engineering Co Ltd in Mumbai is a private company run as a family concern and operating on a small scale.

More detailed information we have received suggests that caution is advisable with this company.

Thank you for treating this advice as strictly confidential.

Yours sincerely

Agency's replies to credit information enquiry

Favourable reply to letter on page 213

Dear

Introduction acknowledges letter and gives initial details —— Thank you for your letter of 10 February regarding A Griffiths & Co.

Details regarding the firm's standing are given with a personal opinion —— We are pleased to report that this is a well-established and highly reputable firm. There are four partners and their capital is estimated to be at least £100,000. They do an excellent trade and are regarded as one of the safest accounts in Cardiff.

Recommendation about credit that could be allowed —— We believe you need not hesitate to allow the initial credit of £12,750 requested. On a quarterly account you could safely allow at least £15,000.

Yours sincerely

Unfavourable reply to letter on page 213

Dear

Introduction acknowledges letter and advises caution —— Thank you for your letter of 10 February regarding A Griffiths & Co. I am sorry to advise caution in their request for credit.

Details are given regarding knowledge of the firm in question —— About a year ago an action was brought against this company by one of its suppliers for recovery of money due, but payment was later recovered in full.

Our enquiries reveal nothing to suggest that the firm is not straightforward. However, the firm's difficulties would seem to be due to bad management and in particular to trading beyond their

The facts as known are stated —— means. Consequently, most of the firm's suppliers either give only very short credit for limited sums or make deliveries on a cash basis.

A reminder that the information should be kept confidential —— This information is supplied in the strictest confidence.

Yours sincerely

Useful expressions

Suppliers' requests for references

- Subject to satisfactory references, we shall be glad to provide the open account facilities requested.
- We hope you will supply the usual trade references so that we can consider open account terms.
- We will be in touch with you as soon as references are received.
- It is our usual practice to request references from new customers, and we hope to receive these soon.

Customers supply references

- Thank you for your letter of ___ in reply to our request for open account terms.
- We have completed and return your credit application form.
- The following firms will be pleased to answer your enquiries . . .
- For the information required please refer to our bankers, who are . . .

Suppliers take up references

- XYZ has given us your name in connection with his (her, their) application for open account terms.
- We have received a large order from ABC, who have given your name as a trade reference.
- We will appreciate any information you can provide.
- Any information provided will be treated in strictest confidence.
- Please accept our thanks in advance for any help you can give us.

Replies to references taken up

- We welcome the opportunity to report favourably on ___.
- Thank you for your letter requesting a reference for ___.
- The firm mentioned in your letter of ___ is not well known to us.
- We would not hesitate to grant this company credit up to £___.
- This information is given to you in confidence and without any responsibility on our part.
- This information is given on the clear understanding that it will be treated confidentially.

15

A typical business transaction (correspondence and documents)

Quotations

Letters of the kind considered in this chapter are handled in business every day. This chapter illustrates their use in a typical transaction in the home trade.

G Wood & Sons have recently opened an electrical goods store in Bristol, and they place an order with Electrical Supplies Ltd, Birmingham, for supplying goods on credit. The transaction opens with a request by G Wood & Sons for information regarding prices and terms for credit.

Request for quotation

G WOOD & SONS

36 Castle Street
Bristol BS1 2BQ
Telephone 0117 954967

GW/ST

15 November 201-

Mr Henry Thomas
Electrical Supplies Ltd
29–31 Broad Street
Birmingham
B1 2HE

Dear Mr Thomas

Thanks for your time on the telephone this morning. As I explained, we have recently opened an electrical goods store on Castle Street Bristol, and have received several enquiries for these domestic appliances:

Swanson Electric Kettles, 2 litre

Cosiwarm Electric Blankets, single-bed size

Regency Electric Toasters

Marlborough Kitchen Wall Clocks

I was glad to hear from you that all these items are available in stock for immediate delivery.

Please let me have your prices and terms for payment two months from date of invoicing. If prices and terms are satisfactory, we would place a first order for 10 of each of these items.

The matter is of some urgency and I look forward to your prompt reply.

Yours sincerely

Gordon Wood
Manager

Supplier's quotation

ELECTRICAL SUPPLIES LTD

29–31 Broad Street Birmingham B1 2HE
Tel: 0121 542 6614

HT/JH

17 November 201-

Mr Gordon Wood
Messrs G Wood & Sons
36 Castle Street
Bristol
BS1 2BQ

Dear Mr Wood

QUOTATION NUMBER E542

Thank you for your enquiry of 15 November. I am pleased to quote as follows:

	£
Swanson Electric Kettles, 2 litre	25.00 each
Cosiwarm Electric Blankets, single-bed size	24.50 each
Regency Electric Toasters	25.50 each
Marlborough Kitchen Wall Clocks	27.50 each

These are current catalogue prices, and we would allow you a trade discount of 33.33%. Prices include packing and delivery to your premises.

It is our usual practice to ask all new customers for trade references. Please let us have the names and addresses of two suppliers with whom you have had regular dealings. Subject to satisfactory replies, we shall be glad to supply the goods and to allow you the two months' credit requested.

As there may be other items in which you are interested, I enclose copies of our current catalogue and price list.

I look forward to working with you.

Yours sincerely

Henry Thomas
Sales Manager

Enc

Request for permission to quote company as reference

A buyer should obtain permission from the suppliers whose names are to be submitted as references. Consent may be obtained verbally if there is urgency, but otherwise the buyer should make this request in writing. Here is one such email request:

Hi Robert

I wish to place an order with Electrical Supplies Ltd, Birmingham, with credit terms. As this will be a first order, they have asked me to supply trade references.

I have been a regular customer of yours for the past four years, and I hope you will allow me to submit your company's name as a reference.

I hope to hear from you soon.

All the best

Gordon Wood
Manager
G Wood & Sons, Bristol

Permission granted

Hi Gordon

Good to hear from you.

During the time we have done business together, you have been a very reliable customer. If your suppliers decide to approach us for a reference we shall be very happy to support your request for credit facilities.

Good luck!

Robert Johnson
Financial Controller
Johnson Traders Ltd

Orders and notes

Order

Covering letter

G WOOD & SONS

36 Castle Street
Bristol BS1 2BQ
Telephone 0117 954967

GW/ST

24 November 201-

Mr Henry Thomas
Electrical Supplies Ltd
29–31 Broad Street
Birmingham
B1 2HE

Dear Mr Thomas

ORDER NUMBER 3241

Thank you for your letter of 17 November quoting for domestic appliances and enclosing copies of your current catalogue and price list.

We have had regular dealings with the following suppliers for the past four or five years. They will be happy to provide the necessary references.

Johnson Traders Ltd, The Hayes, Cardiff CF1 1JW

J Williamson & Co, Southey House, Coventry CV1 5RU

Our order number 3241 is enclosed for the goods mentioned in our original enquiry. They are urgently needed and as they are available from stock we hope you will arrange prompt delivery.

Thank you for allowing two months' credit on receipt of satisfactory references.

Yours sincerely

Gordon Wood
Manager

Enc

Order form

G WOOD & SONS

36 Castle Street
Bristol BS1 2BQ
Telephone 0117 954967

ORDER NO 3241   Date 24 November 201-

Electrical Supplies Ltd
29–31 Broad Street
Birmingham
B1 2HE

Please supply

Quantity	Item(s)	Price
		£
10	Swanson Electric Kettles (2 litre)	25.00 each
10	Cosiwarm Electric Blankets (single-bed size)	24.50 each
10	Regency Electric Toasters	25.50 each
10	Marlborough Kitchen Wall Clocks	27.50 each

Terms   33.33% trade discount

(signature)

for G Wood & Sons

Supplier's acknowledgement

It is good business practice to acknowledge and thank buyers, particularly for a first order and trade reference information. The supplier will then take up the references and put the order in hand when satisfactory replies are received. Here, an order is acknowledged via email:

Dear Mr Wood

Thanks for your order number 3241. We are pleased to confirm that the goods will be supplied at the prices and on the terms stated.

Your order has been passed to our warehouse for immediate despatch of the goods from stock. We hope you will be pleased with them.

Please give me a call if I can be of any further help.

Best wishes

Henry Thomas
Sales Manager
Electrical Supplies Ltd

Advice note

Documents dealing with the despatch and delivery of goods include packing notes, advice of despatch notes, consignment notes and delivery notes. These documents are really copies of the invoice and are often prepared in sets, with the use of NCR (no carbon required) paper, at the same time as the invoice. The copy, which acts as the advice note, will not contain information regarding pricing.

The advice or despatch note informs the buyer that the goods are on the way and enables a check to be made when they arrive. Very often, however, an advice note is replaced either by an invoice sent on or before the day the goods are despatched or sometimes by a letter notifying despatch.

For small items sent by post, a packing note, which is simply a copy of the advice note, would be the only document used. Some suppliers, especially those using their own transport, dispense with the advice note and instead use either a packing note or a delivery note.

Consignment note

When goods are sent by rail, the supplier is required to complete a consignment note representing the contract of carriage with the railway. It gives particulars of the quantity, weight, type and destination of the goods and states whether they are being sent carriage paid (ie paid by the sender) or carriage forward (ie paid by the buyer). In most cases the printed forms supplied by the railway are used but a trader will sometimes prefer to use their own.

The completed consignment note is handed to the carrier when the goods are collected and it travels with them. When the goods are delivered to the buyer the note must be signed as proof of delivery.

Delivery note

Very often two copies of the delivery note are prepared, one to be retained by the buyer, the other to be given back to the carrier signed as evidence that the goods have been delivered. Alternatively the carrier may ask the buyer to sign a delivery book or a delivery sheet recording the calls a carrier has made.

 CAUTION Where it is not possible for the buyer to inspect the goods before signing for them, the signature should be qualified with some comment such as 'not examined' or 'goods unexamined' as a precaution.

Invoices and payments

Invoice

Invoice practice varies. Sometimes the invoice is enclosed with the goods and sometimes it is sent separately, either in advance of the goods (in which case it also serves as an advice note) or after the goods.

 TIP Some organisations now accept a pdf invoice sent via email, others require a hard copy to be mailed. It's always best to check first.

Covering letter

It is not always necessary to send a covering letter with an invoice, but if a letter is sent it need only be very short and formal.

ELECTRICAL SUPPLIES LTD

29–31 Broad Street Birmingham B1 2HE
Tel: 0121 542 6614

HT/JH

3 December 201-

G Wood & Sons
36 Castle Street
Bristol
BS1 2BQ

Dear Sirs

YOUR ORDER NUMBER 3241

We enclose our invoice number 6740 for the domestic electrical appliances supplied to your order dated 24 November.

The goods have been packed in three cases, numbers 78, 79 and 80, and sent to you today by rail, carriage paid. We hope they will reach you promptly and in good condition.

If you settle the account within two months we will allow you to deduct from the amount due a special cash discount of 1.5%.

Yours faithfully

Sally Yap (Mrs)
Credit Control Manager

Enc

Invoice

When G Wood & Sons receive the invoice they will check it with the packing note or delivery note received with the goods to ensure all goods invoiced have been received. They will check the invoice for trade discounts and accuracy before recording it in their books of account.

The invoice may often not be used as a demand for payment but as a record of the transaction and statement of the indebtedness to which it gives rise. The supplier will then later send a statement of account to the buyer.

ELECTRICAL SUPPLIES LTD

29–31 Broad Street Birmingham B1 2HE
Tel: 0121 542 6614

INVOICE

G Wood & Sons
36 Castle Street
Bristol
BS1 2BQ

Date 3 December 201-
Your Order No 3241
Invoice No 6740

For reference purposes the invoice is given a serial number. The order number is also quoted

Quantity	Item(s)	Unit Price £	Total Price £
10	Swanson Electric Kettles (2 litre)	25.00	250.00
10	Cosiwarm Electric Blankets (single-bed size)	24.50	245.00
10	Regency Electric Toasters	25.50	255.00
10	Marlborough Kitchen Wall Clocks	27.50	275.00
			1,025.00
	Less 33.33% trade discount		341.33
			683.67
	VAT @ 17.5%		119.64
			803.31
	3 packing cases (returnable)		15.00
			817.92
	Terms: 1.5% two months		
E & OE	Registered in England No 726549		

The agreed 33.33% trade discount has been given

The terms of payment indicate an allowable cash discount for payment within two months from date of invoice. This discount is deducted at the time of payment

'E & OE' means 'errors and omissions excepted'. It reserves the right for the seller to correct any error in or omissions in the invoice

Debit and credit notes

For the purposes served by these two documents, refer to page 173.

Buyer requests credit note

In our specimen transaction, G Wood & Sons will return the three packing cases charged on the invoice. They will then write to the suppliers asking for a credit note for the invoiced value of the cases. Depending on their usual practice, G Wood & Sons may or may not prepare and send a debit note when making the request.

G WOOD & SONS

36 Castle Street
Bristol BS1 2BQ
Telephone 0117 954967

GW/ST

10 December 201-

Mrs Sally Yap
Credit Control Manager
Electrical Supplies Ltd
29–31 Broad Street
Birmingham
B1 2HE

Dear Mrs Yap

INVOICE NUMBER 6740

We have today returned to you by rail the three packing cases charged on this invoice at a cost of £15.00.

We enclose a debit note for this amount and shall be glad to receive your credit note by return.

All the goods supplied and invoiced reached us in good condition. Thank you for your promptness in dealing with our first order.

Yours sincerely

Gordon Wood
Manager

Enc

▶

◀

G WOOD & SONS

36 Castle Street
Bristol BS1 2BQ
Telephone 0117 954967

DEBIT NOTE

Electrical Supplies Ltd   Date 10 December 201-
29–31 Broad Street
Birmingham
B1 2HE   Debit Note No D 841

Date	Details	Total
		£
10.12.1-	3 packing cases charged on your invoice number 6740 and returned	15.00

Seller issues credit note

When Electrical Supplies Ltd receive the debit note they will check return of the cases. They will then prepare the credit note requested and send it to G Wood & Sons, with or without a covering letter. Any letter sent need only be short and formal, but as this is the buyer's first transaction the supplier would be wise to add a short note to encourage future business.

ELECTRICAL SUPPLIES LTD

29–31 Broad Street Birmingham B1 2HE
Tel: 0121 542 6614

HT/JH

14 December 201-

Mr Gordon Wood
Manager
G Wood & Sons
36 Castle Street
Bristol
BS1 2BQ

Dear Mr Wood

Thank you for your letter of 10 December enclosing debit note number D841.

I confirm receipt of the three packing cases returned. Our credit note number C672 for the sum of £15.00 is enclosed.

Yours sincerely

Sally Yap (Mrs)
Credit Control Manager

Enc

ELECTRICAL SUPPLIES LTD

29–31 Broad Street Birmingham B1 2HE
Tel: 0121 524 6614

CREDIT NOTE

G Wood & Sons
36 Castle Street
Bristol
BS1 2BQ

Date 14 December 201-

Credit Note No C 672

Date	Details	Total
		£
10.12.1-	3 packing cases charged on your invoice number 6740 and returned	15.00

Statement of account

Statements of account are sent to customers at periodic intervals, normally monthly. As well as serving as a request for payment, the statement enables the buyer to compare the account kept by the supplier with the records kept in the buyer's own books. Statements are usually sent without a covering letter.

ELECTRICAL SUPPLIES LTD

29–31 Broad Street Birmingham B1 2HE
Tel: 0121 524 6614

STATEMENT

G Wood & Sons   Date 31 January 201-
36 Castle Street
Bristol
BS1 2BQ

Date	Details	Debit	Credit	Balance
	£	£	£	
3.12.1-	Invoice 6740	818.31		818.31
15.12.1-	Credit note C 672		15.00	803.31
	(2.5% seven days)			
E & OE		Registered in England No 726549		

Payment

Invoices and statements usually indicate the terms of payment. For example:

Prompt cash: A somewhat elastic term but generally taken to mean payment within 15 days from date of invoice or statement.

2.5% 30 days: This means that the debtor is entitled to deduct 2.5% from the amount due if payment is made within 30 days of the invoice or statement, otherwise the full amount becomes payable.

Net 30 days: This means that the debtor must pay in full within 30 days.

Payments in business are usually made by cheque or, if they are numerous, by credit transfer (bank giro). In this transaction the buyer settles the account by sending a cheque to the supplier.

G WOOD & SONS

36 Castle Street
Bristol BS1 2BQ
Telephone 0117 954967

GW/ST

4 February 201-

Mrs Sally Yap
Credit Control Manager
Electrical Supplies Ltd
29–31 Broad Street
Birmingham
B1 2HE

Dear Mrs Yap

We have received your statement of account dated 31 January 201- showing a balance due of £803.31.

From the total amount due on the statement I have deducted the allowable cash discount of 2.5% and enclose a cheque for £783.23 in full settlement.

Yours sincerely

Gordon Wood
Manager

Enc

Receipt

A cheque usually supplies all the evidence of payment necessary, so it is not usual practice for formal receipts to be issued. This does not affect the payer's legal right to request a receipt if one is required.

In the above transaction evidence of payment could be obtained by the supplier's formal receipt or the buyer's cheque after being paid by the bank.

part

5

General Business Correspondence

'Nothing gets in the way of doing business more than language that is anything other than conversational.'

Granville N. Toogood

I n this part we will take a look at a lot of correspondence that takes place within the organisation, ie between you and your internal customers. Yes, I'm calling your colleagues 'customers' because that's exactly what they are. You liaise with people inside your organisation because without their help, you can't get things done, you can't obtain approval on a project, you can't move from point A to point B. So they certainly are your customers, just as you are theirs. So in this part we'll be looking at how you might communicate with your colleagues on a variety of issues.

Just as conflict could happen when you use the wrong tone when speaking to someone, the same could happen when you use the wrong tone in your writing. So with internal correspondence, it's sometimes even more important to make sure you get the tone just right. This is something else we'll be addressing in this part.

With reports being very important and frequently used internal documents, this part looks closely at reports as well as proposals. We'll also look at meetings documentation such as agendas and minutes.

Internal correspondence

The internal memorandum

Before email took over, the most common way to communicate with someone inside your own organisation was using a memorandum. These would be printed messages, sometimes enclosed within reusable envelopes. They would be dropped into employees' in-trays on their desks, or sent within branches of the same company via internal mail.

These days, email is now the usual method for internal communication. However, I do know of some government departments and other organisations where memos are still used, so in this chapter we'll take a look at this document. All the messages in this chapter can of course be used as emails.

Traditional memo headings are To, From, Date and Subject or To, From, Date and Ref. It's not usual to include a salutation or a complimentary close on a memo, but it will usually be signed or initialled.

Memos (or memoranda) serve several purposes:

- to provide information
- to request information
- to inform of actions or decisions
- to request actions or decisions.

Some companies have pre-printed forms for internal memos, but very often templates are saved and used when needed. The typist then only has to insert the relevant details alongside the given headings.

Format

Here is an easy, clear method for displaying internal memos.

Recipient's name and designation

Sender's name and designation

A reference (initials of sender and typist)

No salutation is necessary

Clear subject heading

The body of the memo should be separated into paragraphs, reaching a relevant conclusion and close

No complimentary close is necessary

Leave space for signature (the sender's name and designation are at the top so it is not necessary to repeat these details here)

Enc (if appropriate)

Copy/ies (if appropriate)

MEMORANDUM

To: Hanin Zainomum, Administrative Assistant
From: Clarice Wong, PA to Chairman
Ref: SY/JJ
Date: 14 August 201-

IN-HOUSE DOCUMENT FORMATS

Many congratulations on recently joining the staff in the Chairman's office. I hope you will be very happy here.

I am enclosing a booklet explaining the company's general rules regarding document formats. However, I thought it would be helpful if I summarised the rules for ease of reference.

1. DOCUMENT FORMATS
All documents should be presented in the fully blocked format using open punctuation. Specimen letters, fax messages, memoranda and other documents are included in the booklet. These examples should guide you in our requirements.

2. SIGNATURE BLOCK (LETTERS)
In outgoing letters, it is usual to display the sender's name in capitals and title directly underneath in lower case with initial capitals.

3. NUMBERED ITEMS
In reports and other documents it is often necessary to number items. Subsequent numbering should be decimal, ie 3.1, 3.2, etc.

I hope these guidelines will be useful and that you will study the layouts shown in your booklet. If you have any questions please do not hesitate to ask me.

Clarice Wong

Enc

Copy: HR Department

Memo regarding overdue account

MEMORANDUM

To:	Frank Gates, Financial Director
From:	Michelle Long, Credit Manager
Date:	24 March 201-
Subject:	Overdue account – Carter & Co

Subject heading just like on an email ——

Introduction giving the gist of the information ——

Carter & Co have an overdue account with us for £25,430, despite three reminder letters over the past few weeks.

State the important points ——

At this stage I would normally suggest that you put the matter into our solicitor's hands to recover the debt. However, I know you are a personal friend of Carter's Managing Director, so I wondered if you might wish to write to him to obtain payment of this debt.

Enclose relevant documents ——

Full details are enclosed for your reference.

Action to be taken ——

Please let me know what you decide.

Close ——

Many thanks

Space for signature ——

Enc indicates that something is enclosed ——

Enc

Memo to all staff

MEMORANDUM

To:	All Staff
From:	Anna Vinsen, General Manager
Date:	22 June 201-
Subject:	Emergency Contact Numbers

On several occasions recently there have been some emergencies when it was necessary to contact employees' families. Sometimes it has been very difficult to obtain contact details quickly.

▶

I feel it's important to compile an up-to-date list of contact names and telephone numbers. Please contact Marian Jenner, Personnel Administrator, and let her have at least one name and telephone number (work or home) of someone who we can contact in an emergency.

This information will, of course, be kept confidential.

Many thanks

TIP Note the comma before and after Personnel Administrator. If you want more help on correct use of the comma, please check out Chapter 2.

Internal message following up a phone call

Subject: Penang Orchid complaint by Mark Lim – 10–16 January

Dear John

Intro – thank the caller for his call (don't say 'As spoken' or 'As per') —— Thanks for your call this morning to discuss Mark Lim's complaint about his visit to Penang Orchid. From your enquiries, it seems obvious that conditions were far from satisfactory.

Details – give information in short, simple sentences —— I attach my reply to Mr Lim, from which you will see I've offered him a complimentary one week's holiday for two at any Orchid hotel. I asked him to contact you soon regarding a suitable date.

Action – tell the reader what you want him to do —— When you visit the Penang Orchid next week, please give this matter top priority. Please prepare a detailed report with your recommendations for immediate improvements. Our next regional meeting is on 28 February so I would like your report by 26 February to discuss at the meeting.

Close – when you have asked someone to do something, say 'thank you' —— Thanks for your help.

Sally

CAUTION Avoid old-fashioned phrases or overpoliteness. Say 'Please' instead of 'Would you please' or 'I should be grateful'.

Getting the tone right

Tone means the emotional context of your message, the degree of formality or informality you use in your writing, and the attitude towards the topic and the recipient.

As you most likely know your reader fairly well when writing internally, memos are usually written quite informally. You should aim to put your message over as concisely as possible while still being courteous, clear, concise and correct.

The tone you use in your internal correspondence will be a reflection of these four issues:

- your status
- the reader's status
- your relationship with the reader
- the message you want to give.

Let's look at some sentences written in inappropriate tone, and consider how they can be improved:

I have looked through your report and am totally confused. Please see me urgently to clarify.

I have some questions about your report. Please come and see me to discuss.

Have your report on my desk by 8 am tomorrow at the latest.

Please let me have your report by 8 am tomorrow.

The new price structure you suggested is totally impractical.

I have some suggestions about the new price structure you suggested.

Our phone bills are enormously high. Please tell your staff to stop making so many personal calls.

Our telephone bills have increased considerably. Please ask your staff to avoid non-urgent personal calls during working hours.

IT department can't do anything about your problem. You need a new computer.

 Unfortunately, we cannot fix the issue with your computer. It looks like time to invest in a new one.

 CAUTION Never be unsympathetic, condescending or rude. Always be sincere and clear.

An internal message to a manager where tact is important

Unfriendly and untactful message

This message was written in haste and anger. The manager who reads it would be offended and angry with the sender.

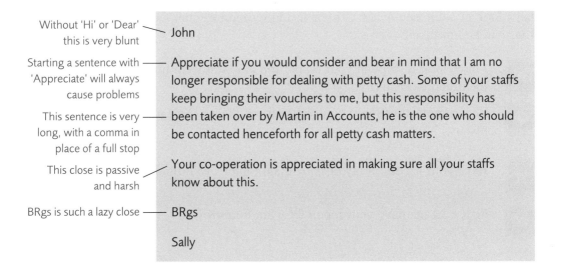

Without 'Hi' or 'Dear' this is very blunt

Starting a sentence with 'Appreciate' will always cause problems

This sentence is very long, with a comma in place of a full stop

This close is passive and harsh

BRgs is such a lazy close

John

Appreciate if you would consider and bear in mind that I am no longer responsible for dealing with petty cash. Some of your staffs keep bringing their vouchers to me, but this responsibility has been taken over by Martin in Accounts, he is the one who should be contacted henceforth for all petty cash matters.

Your co-operation is appreciated in making sure all your staffs know about this.

BRgs

Sally

 CAUTION It's easy for misunderstandings to occur in email. When writing messages, readers could be put on the defensive if your tone is not just right.

The same message in a much more appropriate tone

This message has been written using the four-point structure. The tone is careful and well thought out, considering the senior position of the reader. It would definitely have the desired effect, and it would not cause any offence.

Hi John

Some of your staff are still bringing their petty cash vouchers to me. However, this responsibility has been taken over by Martin in Accounts.

Please let your staff know that they should deal with Martin in future.

Thanks for your help, John.

Sally

Another message that requires good tone

Inappropriate tone

Consider how staff would feel when reading this message. The tone is very poor.

Subject: Lunch hours

It has come to my attention that some employees are taking longer than their scheduled lunch break of one hour.

This situation cannot be allowed to continue. It places an unfair strain on those who try to be on time.

Co-operation of all staff is expected so that this problem can be corrected. Serious repercussions will result for anyone who continues to abuse the system.

Thank you and regards

 CAUTION The third paragraph is written purely in passive voice. For help in using active voice, take a look at Chapter 3.

Much more tactful tone

Subject: Lunch hours

As you know, lunch breaks are scheduled for one hour from 1200 or 1300. Unfortunately, some staff have been extending this one-hour break, and this has caused a lot of inconvenience for others.

I realise there will be times when you may need a longer lunch break. On such occasions, please ensure appropriate cover in advance.

I am sure you understand the importance of keeping to our allocated working hours.

Thank you very much for your co-operation.

 TIP Use words like 'Unfortunately' to soften the tone, and use 'Please' when you are asking the reader to do something.

Using lists and bullets

Lists can be used in any document, and are often used in internal correspondence too. They are useful to set off important points and ideas. You may choose to display the items using numbers, letters or bullets. If there are a lot of points, I prefer to use numbers, then it's much easier to refer to individual numbered points. For a simple list of words or brief points, bullets are better.

Lists are useful for many reasons:

1. They help you to organise your thoughts.
2. They help focus your reader's attention on important points.
3. They help readers find key points.
4. They simplify detailed or complicated topics.
5. They make it easier for readers to skim.
6. They enhance visual impact.

When compiling a list, it's important to make sure your listed items are parallel in structure. For example, if one item begins with a verb, make sure they all do, as shown here:

Non-parallel list items	Parallel list items
Your document must look attractive.	Make your document look attractive.
Diagrams will be visually appealing.	Use diagrams for a visual appeal.
Too many fonts make it hard to focus.	Avoid using too many different fonts.
Layout must be consistent.	Be consistent in your layout.
Courtesy and tact are important.	Use courtesy and tact in your writing.

A numbered list

Note that all these points begin with active verbs.

11 GOLDEN RULES FOR BUSINESS EMAIL ETIQUETTE

1. Reply to all emails promptly.
2. Use 'Reply all' with caution and only when essential.
3. Down-edit your replies by removing anything that's not necessary.
4. Include a SMART subject line on every message.
5. Start with a proper greeting and appropriate sign-off.
6. Ensure email addresses are correct.
7. Do not type messages in ALL CAPS.
8. Write in full sentences and correct grammar.
9. Use contractions occasionally, but not abbreviations.
10. Format messages neatly, with a space between paragraphs.
11. Structure messages logically using Shirley's four-point plan.

 TIP Your colleagues, your superiors and your peers will judge you on the quality of your written communication. Make sure you give a good impression.

A message including a simple bulleted list

Subject: Internal training

Hi Martin

Thanks for enquiring about internal training programmes.

We have several that I know you'll be interested in. Here are the upcoming dates of some of our most popular programmes:

- Powerful Presentations – 12–13 June
- Grammar for Professionals – 17 June
- How to Lead a Team – 22 July
- Essential Communication Skills – 21–22 August
- Organise your Business Life Now – 24 August
- Assertiveness and You – 21 September

I suggest you register early for any of these programmes. You can access the online registration form on our Intranet site.

If you have any questions please call me on extension 246.

Susan

Internal email to staff including a two-column list

Subject: 10th anniversary celebration – 10 May

It has been a prosperous journey for ST Training Solutions this past decade. To celebrate our 10th anniversary we have arranged a special evening, and we would like you to share this with us.

We promise a night of glamour and humour – with Radio DJ, The Flying Scotsman, as our MC, and an enchanting and glamorous performance by local singing sensation, Clarissa Monteiro.

Details for your diary are:

Date   Friday 10 May 201-
Time   Cocktails from 1900; Dinner at 1930; Showtime 2030
Venue   Ritz Carlton Hotel, Fullerton Road
Theme   Starry Starry Nite!

Please confirm your attendance to Eunice at eunice_soon@gic.com.sg by 9 April.

We have mystery gifts for early birds and prizes for the best dressed star! Come join us and be among the stars for the night!

Let's celebrate our 10th anniversary in style!

Another simple list of bullets that are parallel in structure

In this example, note that every bullet must follow ' . . . by:'.

You can improve your business writing by:

- adopting a friendly, conversational writing style
- reading your message out loud to check the tone
- keeping to the point and staying focused
- organising your points carefully and logically
- using language that the reader will understand.

An internal memo to staff including numbered points

MEMORANDUM

To: All Staff
From: John Staples, Security Manager
Date: 29 September 201-
Subject: New car parking arrangements

Our new car parking arrangements will come into effect on Monday 14 October.

A parking plan is attached showing special areas in the car park for staff and visitors. A copy has also been placed on all company notice boards.

Procedures for car parking are:

1. **Staff**
   If you have a car, please collect your *red permit* from the security gate, and display this on your car windscreen. Please park only in areas specially reserved for staff.

2. **Visitors**
   Special areas have been reserved for visitors. Please inform security when you are expecting visitors. *Green permits* will be given to them when they report to the security gate.

Unauthorised parking will lead to many problems for the large lorries that deliver raw materials to our factory.

Thank you for your help in ensuring that these new car parking arrangements are successful.

 TIP Prepare your internal correspondence with as much care as external messages. Internal staff are your customers too.

Internal does not mean shoddy

Many people mistakenly think that because they are writing to their colleagues, not clients, they can type out internal messages quickly and not bother to proofread. As a result, very often such messages are filled with typos, non-sentences or very long-winded sentences that just ramble on. Consider the consequences of this action:

- The reader has to read the message several times to try to understand it.
- The reader shoots back a message asking for clarification.
- The originator replies again in a rush and creates more confusion.
- The reader shoots back another message asking for clarification.
- This ding-dong continues, wasting everyone's time.

We are now in the era of rushed typing, even thumb-typing on handheld devices. Your reader doesn't have time to waste, and neither do you. You need to be accurate, clear and concise, so that you can get your message across effectively.

 TIP Everyone is busy. Show some respect for your colleagues by making sure your internal messages are clear and concise, and by proofreading carefully before you hit 'send'.

 ## Checklist

- Take as much care with internal correspondence as you do with external.

- Keep your messages brief, concise and clear, and do check through carefully before sending.

- Consider your status, the reader's status, your relationship and your message when composing any document.

- Use the right tone in all messages, so that the reader isn't offended or hurt.

- Don't write messages in haste, or they will probably come back to haunt you.

- Use numbered points or lists to help you and the reader focus on important points.

- Consider sub-headings when the message contains more than one related topic.

- Remember grammatical parallelism when constructing any list.

- Read your messages out loud to consider the tone you would use if you were speaking.

- Craft all your messages in line with the four-point plan so that they are structured logically.

17

Secretarial and administrative correspondence

In any organisation, the administrative or secretarial staff play an important role. The Secretary or PA, often called an Executive Assistant, will usually deal with lots of correspondence regarding appointments, meetings, interviews, conferences, travel and accommodation.

In this chapter we will look at some of the correspondence that is typically dealt with by today's secretarial or administrative support staff.

Arranging meetings

Email requesting appointment

Request

> Dear Mr Harrison
>
> Our Mr Chapman has informed me that you have returned home from your visit to the Middle East. There are a number of points that have arisen on the book I am writing on Modern Business Organisation. I would like the opportunity to discuss these with you.
>
> I shall be in London from 16 to 19 September. Do you have a date and time during this period when we could meet?
>
> I look forward to seeing you again.

Reply

> Dear Mr Alexander
>
> Great to hear from you.
>
> I will look forward to meeting you again to discuss this. I am available any time on 17 or 18 September afternoon. When you arrive in UK, please give me a call on +020 8246 4172 and we can make arrangements.
>
> See you soon.

Email requesting appointment

Enquiry

> Dear Mr Jones
>
> I am concerned about the difficulties you are having with the goods we supplied earlier this year.
>
> I shall be in your area next week and would very much like the opportunity to discuss this matter with you personally. Do you have some time to meet me on 13 or 14 September?
>
> Many thanks

Reply

> Dear Mrs Graham
>
> Thanks for your email.
>
> I would be very pleased to meet you at my office on 14 September. Is 2.30 pm good for you?
>
> I look forward to seeing you.

Email regarding board meeting

Dear All

I confirm that there will be a board meeting on Monday 5 April 201-.

The scheduled board meeting dates for the remainder of 201- are:

Wednesday 5 May

Wednesday 7 July

Wednesday 1 September

Wednesday 3 November

All meetings will start at 1000 and will be held in the board room in Atrium Towers. Lunch will follow the meetings, hosted by the Chairman, Mr Graham Newman.

Please inform your directors of these arrangements.

Many thanks

Mary

 CAUTION So many people would begin this email with 'Please be informed' or 'Please be advised'. This is passive and outdated. To learn more about using active voice, see Chapter 3.

Email regarding diary dates

Email suggesting dates

Good morning Ladies

The Chairman has decided to hold an extra board meeting in August to discuss urgent issues. The available dates are:

Tuesday 10 August

Thursday 12 August

Friday 13 August

Timings will be 1100 to 1400 or 1400 to 1700.

Please let me know which of these dates will be good for your boss.

Best

Joy

TIP I prefer displaying the date as day/date/month/year, ie Tuesday 10 August 201-. If you prefer to put the comma after the day, that's also fine. You may prefer to use Tuesday August 10th, 201-. This is also fine. Just be consistent. See more about this topic in Chapter 5.

Reply

Hi Joy

The Finance Director will be available on Thursday 12 or Friday 13 August. His diary is free all day on both these dates.

Cheers

Sandra

 TIP Note how these emails are finished with 'Best' and 'Cheers'. These are great alternatives to 'Regards', which I feel is overused. Why not come up with your very own closing? Read more about closings on emails in Chapter 6.

Arranging events and speakers

Letter inviting speaker to conference

Invitation

Dear Miss Forrester

Mention function, location, dates, number of delegates expected —— Our Society will be holding a conference at the Moat House Hotel, Swansea from 4 to 6 October with the theme 'The 21st Century Secretary'. Approximately 100 delegates are expected, comprising mostly practising secretaries, PAs, office managers, plus some lecturer members.

Include title of talk and timing. Mention any payments to be made —— I hope you will agree to speak on the subject of 'Building Great Relationships' on 5 October from 1030 to 1130. We would be prepared to pay you the usual fee of £500 and your travel expenses. We will also arrange hotel accommodation for you for one night on 4 October.

Enclose detailed —— The detailed draft programme is enclosed. We hope you will also
programme   stay on to attend other sessions of the conference too.

Request confirmation —— We really hope you can accept our invitation. At the same time
and details of any   please let us know what equipment you will need.
equipment needed

I hope to hear from you soon.

Yours sincerely

 TIP Note how this message is structured in accordance with the four-point plan – introduction, details, action, close. This is a great formula to follow. Learn more in Chapter 4.

Programme

THE 21st CENTURY SECRETARY

DRAFT PROGRAMME

DAY ONE – MONDAY 4 OCTOBER 201-

0800	Registration and morning coffee
0900	Chairman's opening remarks (Hayati Abdulla, Core Services)
0915	The new secretary (Sally Turner, Author)
1030	Refreshments
1100	Business writing skills (Janice Lim, STTS Training Consultancy)
1230	Lunch
1400	Working with different cultures (Nigel Lau, StaSearch International)
1530	Refreshments
1600	Convincing presentation skills (Robert Ling, Maxim Training)
1715	End of Day One

▶

◀

DAY TWO – TUESDAY 5 OCTOBER 201-

0800	Morning coffee
0900	Chairman's opening remarks (Janice Lim, STTS Training Consultancy)
0915	IT and the new economy (Sarah Cowles, D&P International)
1000	Refreshments
1030	Building great relationships (Pamela Forrester)
1130	Lunch and fashion show
1400	Effective time management (Ian Norton, Hardwick Industries)
1515	Refreshments
1545	Projecting the right image (Sharon Conway, C&G Fashion House)
1630	How to be a super secretary (Sally Turner, Author)
1715	End of Day Two

Reply accepting invitation

Dear Ms Bolan

Thank you —— Thank you for your letter inviting me to speak at your conference on 5 October on the subject of 'Building Great Relationships'.

Acceptance and —— I am delighted to accept your invitation, and confirm that I shall confirmation require overnight accommodation on 4 October.

Mention equipment —— I shall bring my laptop to conduct my presentation, so all I will needed need you to arrange is the usual projector.

An appropriate close —— I look forward to meeting you and other members of your Society again at your conference.

Best wishes

Letter declining invitation

Dear Mr Woodhead

Many thanks for your email with details of your upcoming conference.

I am sorry to say that I have an overseas commitment in October, so unfortunately I will not be able to speak at your conference.

I am sure the day will be a great success.

Email regarding conference accommodation

Dear Sirs

Mention function, date, timing and reason for writing — Our company will be holding a one-day conference on Saturday 18 May from 1000 to 1730, and we are looking for a suitable venue for 200 delegates.

Include number of delegates expected

Our requirements are:

Number and list the specific requirements —
1. A conference room with theatre-style seating
2. A foyer for table top exhibitors to display products/brochures
3. A reception area for welcoming and registering delegates
4. Morning refreshments at 1100 and afternoon refreshments at 1530
5. A buffet lunch to be served from 1300 to 1400

Ask for information about facilities and costs — If you have suitable facilities available please let us know the costs involved. At the same time, please send specimen menus for the buffet lunch as well as refreshment choices.

We hope to hear from you soon.

Many thanks

Conference programme

A programme is a list of things happening at an event. The items will be displayed in column format in chronological order.

Company's name —— ① **Turner Communications**

Main heading —— **5TH ANNIVERSARY CELEBRATIONS**

Date and venue —— to be held on Wednesday 17 September 201-
at Supreme Hotel, Aston, Sheffield

Sub-heading states
whether
programme is
provisional or final
—— PROVISIONAL PROGRAMME

1800	Arrival of directors and staff
1830	Arrival of guests
	5th Anniversary folders will be issued to guests on arrival
	Cocktails will be served
1900	Introduction by Suzanne Sutcliffe, Marketing Manager, who will act as Master of Ceremonies
1915	Opening address (Sally Turner, Managing Director)
1930	Slide presentation (Mandy Lim, Administration Manager)
2000	Buffet supper
2130	Toastmaster (John Stevens, Public Relations Manager)
2145	Closing address (Suzanne Sutcliffe, Marketing Manager)

Use 24-hour clock, with
or without the
word 'hours'

List each item in turn
with any extra
details

Include names of
presenters in brackets

Clarify the time at which —— Drinks will be served until 2300
the event will finish

SS/ST

5 July 201-

TIP For some good websites for secretaries, check out:

www.deskdemon.com

www.executivepa.com

www.pa-assist.com

www.executivesecretary.com

www.iaap-hq.org

Invitations and replies

Many companies organise special functions to publicise certain events, for example:

- the opening of a new branch office
- the introduction of new products or services
- the retirement of a senior executive
- a special anniversary

Formal invitations are usually printed on A5 or A6 high quality paper or card. RSVP at the foot means 'Please reply' or 'Respondez s'il vous plait'.

Formal invitation

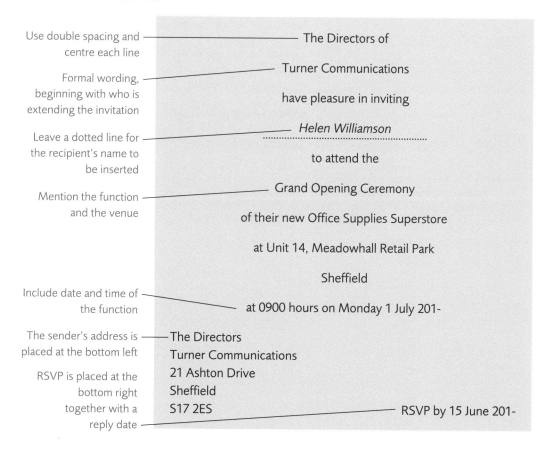

Use double spacing and centre each line — The Directors of

Formal wording, beginning with who is extending the invitation — Turner Communications

have pleasure in inviting

Leave a dotted line for the recipient's name to be inserted — *Helen Williamson*

to attend the

Mention the function and the venue — Grand Opening Ceremony

of their new Office Supplies Superstore

at Unit 14, Meadowhall Retail Park

Sheffield

Include date and time of the function — at 0900 hours on Monday 1 July 201-

The sender's address is placed at the bottom left — The Directors
Turner Communications
21 Ashton Drive
Sheffield
S17 2ES

RSVP is placed at the bottom right together with a reply date — RSVP by 15 June 201-

When accepting or refusing an invitation it is usual to do so in a similar style to the invitation that was received. If the invitation is refused it is courteous to give a reason.

Reply to formal invitation

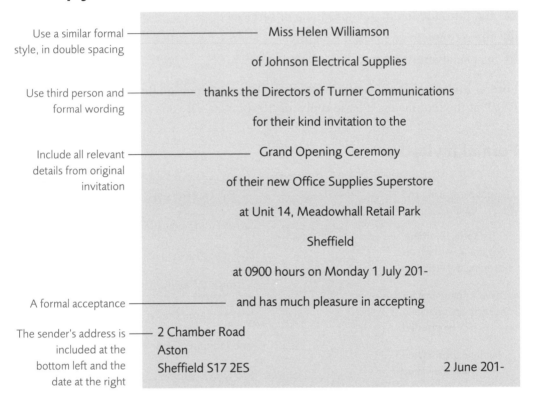

Use a similar formal style, in double spacing — Miss Helen Williamson

of Johnson Electrical Supplies

Use third person and formal wording — thanks the Directors of Turner Communications

for their kind invitation to the

Include all relevant details from original invitation — Grand Opening Ceremony

of their new Office Supplies Superstore

at Unit 14, Meadowhall Retail Park

Sheffield

at 0900 hours on Monday 1 July 201-

A formal acceptance — and has much pleasure in accepting

The sender's address is included at the bottom left and the date at the right — 2 Chamber Road
Aston
Sheffield S17 2ES

2 June 201-

Email invitation

Invitation

Hi Shirley

I hope you remember meeting me in Jakarta earlier this year when you conducted a two-day workshop for Marshalls employees.

I was interested to see from your e-newsletter that you will be in Jakarta next month conducting some more workshops. My company will be celebrating its 10th anniversary by holding a dinner at the Aryaduta Hotel on Wednesday 29 August.

It would be great if you could join us at our special celebration. Please let me know as soon as possible so that I can add your name to our VIP list.

Best wishes

Sally Turner
HR Manager
Marshalls Indonesia Sdn Bhd

Reply to email invitation accepting

Hi Sally

Of course I remember meeting you. I really enjoyed my workshop with Marshalls and hope you have seen your photographs in my photo gallery.

Thank you so much for your kind invitation to join you at your company's 10th anniversary dinner on Wednesday 29 August. I should normally be flying into Jakarta that evening for some public workshops later that week. However, I am going to change my travel plans and fly in one day early so that I can accept your invitation.

I will certainly look forward to seeing you and your colleagues again. Thank you so much for thinking of me.

Shirley
www.shirleytaylortraining.com

Reply to email invitation declining

Hi Sally

I do indeed remember meeting you. I really enjoyed my workshop at Marshalls and hope you have seen your photographs in my photo gallery.

Thank you so much for your kind invitation to join you at your company's 10th anniversary dinner on Wednesday 29 August. However, I shall only be in Jakarta from Sunday 17 for one night – I have to fly back to Singapore straight after my workshop due to commitments in Singapore.

I am so sorry that I cannot join you on this wonderful evening. Please pass on my apologies to all my friends at Marshalls.

I hope to see you again when I am back in Jakarta for a longer trip.

My best wishes

Shirley
www.shirleytaylortraining.com

Another example of an invitation

Here is an invitation sent out to invite clients to a special afternoon presentation.

Rendezvous with Shirley

How would you like to transform your business writing style in just five simple steps? Learn modern language instead of yesterday's jargon? Overcome common problems in today's business writing? Learn the art of writing as you speak?

Imagine all this in just two hours!

What's more amazing is that it will cost you nothing except your time. If that's not good enough for you, we will even give you a free book – Shirley Taylor's *Model Business Letters, Emails and Other Business Documents* seventh edition.

'Rendezvous with Shirley' is our way of getting great minds to converge in a pleasant setting, and we will all learn a thing or two as well. We can also network with peers in an afternoon of learning fun.

Event	Rendezvous with Shirley
Topic	Make writing your most powerful tool
Venue	Room 1, STTS Training Centre, Wisma Atrium
Date	Friday 17 March 201-
Time	1500–1700

Please reserve your place early by calling me on 65323414.

See you soon.

Itineraries

An itinerary gives full details of a journey in order of date. It shows all travel arrangements, accommodation and appointments. It is usual to use sub-headings and columns so that the information is displayed attractively and is easy to refer to.

Use plain paper and show the company's name

Turner Communications

ITINERARY FOR MRS SALLY TURNER

Include traveller's name, places being visited and duration of trip

TOUR OF SINGAPORE AND MALAYSIA

7–19 JULY 201-

Display all dates as shoulder headings

SUNDAY 7 JULY

| 1530 | Depart London Heathrow (flight SQ101) |

MONDAY 8 JULY

Use a two- or three-column format for ease of reference

| 1830 | Arrive Singapore Changi Airport (Met by Scott Milnes, Communications Asia) Accommodation: Supreme International Hotel, Scotts Road. |

TUESDAY 9 JULY

| 1030 | Miss Joy Chan, Communications Asia, Funan Centre |
| 1430 | Mr Andrew Walton, TalkTime, Bugis Junction |

WEDNESDAY 10 JULY

Use 24-hour clock for all times

| 0930–1730 | 5th International Telecommunications Conference |

SUNDAY 14 JULY

| 1545 | Depart Singapore Changi Airport Terminal 2 (flight MH989) |
| 1700 | Arrive Kuala Lumpur Accommodation: Royal Hotel, Petaling Jaya |

MONDAY 15 JULY

| 1030 | Mr Keith Walker, KL Talk |
| 1530 | Mrs Ong Lee Fong, Malaysia Communications |

TUESDAY 16 JULY

| 1130 | Miss Sylvia Koh, Talklines |

FRIDAY 19 JULY

| 2330 | Depart Kuala Lumpur (BA 012) |

SATURDAY 20 JULY

| 0830 | Arrive London Heathrow |

ST/BT

Reference and date

15 June 201-

Arranging accommodation

Booking company accommodation at a hotel

In this enquiry a company writes to the manager of a London hotel requesting information about accommodation.

Email enquiry

> Dear Sirs
>
> My company will be displaying products at the forthcoming British Industrial Fair at Earls Court. We will require hotel accommodation from 13 to 17 May inclusive.
>
> Please let me know if you can extend special terms to us. Please also indicate if you have one double and three single rooms available during this period.
>
> I hope to hear from you soon.
>
> Name

Reply

> Dear Miss Johnson
>
> Thank you for your email.
>
> We do have one double and three single rooms available from 13 to Friday 17 May. We will offer you our special Internet rate, which is £90 for the double room and £80 for each single room. However, as we are now entering the busy season and bookings for this period are likely to be heavy, we suggest that you make your reservation without delay.
>
> You can learn more about us at our website **www.courtcrest.co.uk**. It is possible to reach Earls Court by public transport within 15 minutes, or by car in 8 minutes.
>
> I hope to receive confirmation of your reservation soon.
>
> Name

Confirmation of reservation

In the first instance you may telephone the hotel to make your reservation. This would be confirmed in writing immediately.

Dear Mr Nelson

Thank you for your email and our telephone conversation today.

I confirm reservation of one double and three single rooms from 13 to 17 May inclusive. Names of guests are:

Mr & Mrs Philip Andersen

Mr Geoffrey Richardson

Miss Lesley Nunn

Mr Jonathan Denby

The account will be settled by Mr Philip Andersen, our Company's General Manager.

Many thanks

Asking for sponsorship

Very often secretarial staff have to organise an event like the company's annual dinner and dance. If sponsorship is needed, the secretary may have to write a circular to many organisations asking for contributions. Here is one such letter.

Letter asking for sponsorship

Subject: ST Training Solutions Annual Dinner and Dance – 10 December 201-.

ST Training Solutions is celebrating its 10th Anniversary on 10 December 201-. We are holding a special event at the Fullerton Hotel to commemorate the occasion with our staff and guests.

In order to make this event special and successful, we are looking for contributions from our valued business partners like you. We hope you can donate gifts or vouchers to be given out to prize winners as door gifts, lucky draw prizes or games prizes.

Please complete the attached contribution form and fax it back to me at 6722 0739.

If you have any questions, please give me a call at 62348288.

I look forward to a positive reply from you soon.

Many thanks

Meetings documentation

Many meetings take place in business, and an effective meeting is an efficient tool in the communication process. Meetings provide a useful opportunity for sharing information, making suggestions and proposals, taking decisions and obtaining instant feedback.

While meetings are becoming less formal, it's still important for the appropriate paperwork to be dealt with and for the minutes to be recorded accurately.

Many meetings may be held in business, for example:

- A staff or team meeting, in which team members or junior staff discuss certain issues with the manager or the immediate superior.
- A special purpose meeting, for example a team set up to organise the company's annual dinner and dance.
- A management meeting or, if they are all directors of the organisation, a board meeting.

Notice and agenda

The success of any meeting depends on essential preparatory work. Part of this work involves making sure that all the documentation is in order. The notice and agenda are usually combined in one document. The portion at the top is known as the notice – this gives details of the type, place, day, date and time of the meeting. The agenda is the list of topics to be discussed at the meeting. A well-compiled agenda will help focus the direction of a meeting and ensure all the necessary discussion points are covered within the allotted time period.

Compiling the agenda

1 It is usually the secretary who compiles the agenda, in liaison with the Chairman. The first step is to gather together all relevant information, sort out the items for discussion and then consider the detail that needs to be covered in the meeting. It's always a good idea to consider the time each item is likely to take, and actually assign a time limit to each item. By doing this, you can ensure the overall time allocated to the meeting is appropriate.

2 An agenda will always be headed with the date, time and location of the meeting, and restricted to one sheet of paper if possible. The traditional structure is shown here:

- Welcome and introductions (if necessary)
- Apologies for absence
- Approve minutes of last meeting
- Matters arising from last meeting
- List of issues to be discussed
- Any other business
- Details of next meeting

It's usual to number each item, and leave plenty of room in the margins for notes.

3 Housekeeping matters, such as apologies for absence and approval of the last meeting's minutes, should be placed at the top, followed by regular reports from members. Topics should then be ordered on the agenda logically, and items with a similar theme grouped together. This will reduce the risk of revisiting the same ground over and over. Begin with routine and straightforward business where decisions are likely to be easy and uncontroversial.

Once the less contentious issues are out of the way, current issues should be listed, and this is normally where the bulk of discussion will be. Be generous in the time you allow for discussion of these agenda items.

Finally, allow for any other business and plan to set the date, time and location of the next meeting.

Circulating the agenda

It's always good practice to circulate a preliminary agenda well in advance and ask for feedback and any additional items from members. Circulate the final agenda as far in advance as possible, and attach any relevant papers to allow those attending enough time to prepare. Include an indication of the likely duration of the meeting overall and the time allocation dedicated to each individual item. If in doubt, be generous with timings: participants are happy for meetings to finish earlier than planned but not so happy if they overrun.

Email requesting agenda items

Subject line states name, ——— **Subject:**   Operations Meeting 14 July 1030
date and time of meeting

Clarify meeting, venue, ——— The next Operations Meeting will be held in the Conference
time and date of meeting   Room at 1030 on Monday 14 July.

New items for the agenda are currently:

Mention any items already ——— ▪ New brochure (Suzanne Sutcliffe)
included on the agenda   ▪ Annual Dinner and Dance (Mandy Lim)

Give a deadline for ——— If you wish to add any further items to the agenda please let me
submitting extra items   know before 8 July.

Many thanks

CAUTION Note that the first line does not begin 'Please be informed'
or 'Please be advised'. These phrases are passive as well as outdated.
Learn more about getting rid of redundant phrases like these and using
active voice in Chapter 3.

Email including agenda

Subject: Operations Meeting 14 July 1030

Hi all

Confirm details regarding ——— The next monthly Operations Meeting will be held in the
venue, date and time   Conference Room at 1030 hours on Monday 14 July.

The agenda will be:

These first three items ——— 1. Apologies for absence
of 'ordinary business'
should be included on   2. Minutes of last meeting
every agenda   3. Matters arising from the minutes

4. New brochure (Suzanne Sutcliffe)

◄

These are items of 'special
business', specific to this
meeting only
| 5. Annual Dinner and Dance (Mandy Lim)
| 6. New branches (Doug Cowles)
| 7. Far East Trip (Sally Yap)
| 8. European Telecommunications Conference (John Stevens)
| 9. 5th Anniversary Celebrations (Mark Lim)

These final two items are
again 'ordinary business'
| 10. Any other business
| 11. Date of next meeting

CAUTION When you use the automatic numbering in your Word
program, you'll find that it does not automatically align the tenth point
neatly. I always change this because it looks so much better when all
the numbers are aligned.

Notice and agenda

Company's name — ②**Turner Communications**

Title of meeting — **OPERATIONS MEETING**

Notice section: state
venue, time and date
The monthly Operations Meeting will be held in the Conference
Room at 1030 on Monday 14 July 201-

AGENDA

Opening ordinary
business
1. Apologies for absence
2. Minutes of last meeting
3. Matters arising from the Minutes

Special business (note
full names in brackets)
4. New brochure (Suzanne Sutcliffe)
5. Annual Dinner and Dance (Mandy Lim)
6. New branches (Doug Cowles)
7. Far East Trip (Sally Yap)
8. European Telecommunications Conference (John Stevens)
9. 5th Anniversary Celebrations (Mark Lim)

Final ordinary business — 10. Any other business
11. Date of next meeting

ST/BT

TIP Very often it's the secretary's job to draw up the agenda for a meeting. You can see more secretarial and administrative correspondence in Chapter 17.

Minutes of meeting

Minutes are a written record of what took place at a meeting. An accurate written record is essential not only for those who attend the meeting but also for those who were absent. The purpose of taking minutes is to keep an accurate record of events for future possible reference. Minutes should record:

- when the meeting took place
- who was in attendance
- who was absent
- what was discussed
- what decisions were made

Meeting minutes are a record of discussions and decisions, and over time they might form an important historical record.

 TIP Always use past tense for minutes, with third person and reported speech.

Types of minutes

Verbatim minutes

These are used primarily in court reporting, where everything needs to be recorded word for word.

Minutes of resolution

Only the main conclusions that are reached are recorded, not a note of the discussions that took place. These are usually used for minutes of Annual General Meetings and other statutory meetings. It is important to note the exact wording of any resolutions that are passed. For example:

PURCHASE OF PHOTOCOPIER

The Company Secretary submitted a report from the Administration Manager containing full details of the trial of the AEZ photocopier.

IT WAS RESOLVED THAT the AEZ photocopier be purchased at a cost of £11,500.

Minutes of narration

These minutes are a concise summary of all the discussions that took place, reports received, decisions made and action to be taken.

PURCHASE OF PHOTOCOPIER

The Company Secretary submitted a report from the Administration Manager containing full details of the trial of the AEZ photocopier. The machine had been used for a period of four weeks in the Printing Room. Its many benefits were pointed out, including reduction/enlarging features and collating. After discussion it was agreed that such a machine would be extremely valuable to the company.

The Company Secretary was asked to make the necessary arrangements for the photocopier to be purchased at the quoted price of £11,500.

The amount of detail recorded in minutes will depend on the type of meeting and maybe its historical culture. Some organisations like to have a record that records the essence of the discussions that took place. Others just want to record the decisions that are made. You will need to check back on previous minutes to see how much detail is required.

Minutes of meeting

AURORA HOLDINGS plc

WELFARE COMMITTEE

MINUTES OF A MEETING OF THE WELFARE COMMITTEE HELD IN THE CHAIRMAN'S OFFICE ON TUESDAY 21 OCTOBER 201- AT 1630.

PRESENT: Eileen Taylor (Chairman)

        Jim Cage

        Robert Fish

        Ellen McBain

        Wendy Sheppard

        Georgia Thomas

        Will Thomas

1. APOLOGIES FOR ABSENCE

   Apologies were received from Anthony Long who was attending a business conference.

2. MINUTES OF LAST MEETING

   The minutes had already been circulated and the Chairman signed them as a correct record.

3. MATTERS ARISING

   Will Thomas reported that he and Georgia had visited Reneé Simpson in hospital on 16 October to deliver the committee's basket of flowers and good wishes for a speedy recovery. Reneé said she hopes to return to work on Monday 4 November and will be able to attend the next committee meeting.

4. STAFF RESTAURANT

   Jim Cage distributed copies of the accounts for the half year ending 31 July. He pointed out that a profit of £1,300 was made over the first six months of the year. He suggested that some of this be used to buy a new coffee machine as the present one is old and unreliable. It was agreed that he would obtain some estimates and discuss this further at the next meeting.

5. WASHROOM FACILITIES

   Mr Taylor announced that several complaints had been received about the female toilets on the second floor. He had investigated the complaints and agreed that they

▶

need upgrading. Several locks were reported to be faulty, plus chipped tiles and poor decoration.

Miss McBain volunteered to arrange for some local workmen to provide an estimate on the cost of repairs and to report back at the next meeting.

6. STUDY LEAVE FOR YOUNG TRAINEES

Mr Robert Fish reported that examinations would be held in December for the company's trainees who presently attend evening courses at Cliff College. He suggested that they should be allowed two weeks' study leave prior to their examination.

The Chairman pointed out that it was not within the committee's power to make this decision. She advised Mr Fish to write formally to the Board of Directors asking them to include this item on the agenda of the November Board Meeting. An answer should be obtained before the next meeting.

7. CHRISTMAS DINNER AND DANCE

Miss Wendy Sheppard passed around sample menus that had been obtained from hotels. After discussion it was agreed that arrangements should be made with the Marina Hotel for Saturday 21 December. Miss Sheppard agreed to make all the necessary arrangements.

8. ANY OTHER BUSINESS

There was no other business.

9. DATE OF NEXT MEETING

It was agreed that the next meeting would be held on Wednesday 20 November at 2000.

(Chairman)

(Date)

ET/ST

30 June 201-

TIP Remember to use past tense and reported speech in minutes:

was	*not*	is
would be	*not*	will be
had been	*not*	has been
were	*not*	are

Meetings terminology

(This list is reproduced with permission of Desk Demon, the UK's No 1 spot for secretarial resources, information and community, from their website **www. deskdemon.co.uk**).

Ad hoc: from Latin, meaning 'for the purpose of', as for example, when a sub-committee is set up specially to organise a works outing

Adjourn: to hold a meeting over until a later date

Adopt minutes: minutes are 'adopted' when accepted by members and signed up by the chairman

Advisory: providing advice or suggestion, not taking action

Agenda: a schedule of items drawn up for discussion at a meeting

AGM: Annual General Meeting: all members are usually eligible to attend

Apologies: excuses given in advance for inability to attend a meeting

Articles of Association: rules required by Company law which govern a company's activities

Attendance list: in some committees a list is passed round to be signed as a record of attendance

Bye-laws: rules regulating an organisation's activities

Casting vote: by convention, some committee chairmen may use a 'casting vote' to reach a decision, if votes are equally divided

Chairman: leader or person given authority to conduct a meeting

Chairman's Agenda: based upon the committee agenda, but containing explanatory notes

Collective Responsibility: a convention by which all committee members agree to abide by a majority decision

Committee: a group of people usually elected or appointed who meet to conduct agreed business and report to a senior body

Consensus: agreement by general consent, no formal vote being taken

Constitution: set of rules governing activities of voluntary bodies

Convene: to call a meeting

Decision: resolution minutes are sometimes called 'decision minutes'

Eject: remove someone (by force if necessary) from a meeting

Executive: having the power to act upon taken decisions

Extraordinary Meeting: a meeting called for all members to discuss a serious issue affecting all is called an Extraordinary General Meeting; otherwise a non-routine meeting called for a specific purpose

Ex officio: given powers or rights by reason of office

Guillotine: cut short a debate – usually in Parliament

Honorary post: a duty performed without payment, eg Honorary Secretary

Information, Point of: the drawing of attention in a meeting to a relevant item of fact

Intra vires: within the power of the committee or meeting to discuss, carry out

Lie on the table: leave item to be considered instead at the next meeting (see Table)

Lobbying: a practice of seeking members' support before a meeting

Minutes: the written record of a meeting; resolution minutes record only decision reached, while narrative minutes provide a record of the decision-making process

Motion: the name given to a 'proposal' when it is being discussed at a meeting

Mover: one who speaks on behalf of a motion

Nem con: from Latin, literally, 'no one speaking against'

Opposer: one who speaks against a motion

Order, point of: the drawing of attention to a breach of rules or procedures

Other business: either items left over from a previous meeting, or items discussed after the main business of a meeting

Point of order: proceedings may be interrupted on a 'point of order' if procedures or rules are not being kept to in a meeting

Proposal: the name given to a submitted item for discussion (usually written) before a meeting takes place

Proxy: literally 'on behalf of another person' – proxy vote

Quorum: the number of people needed to be in attendance for a meeting to be legitimate and so commence

Refer back: to pass an item back for further consideration

Resolution: the name given to a 'motion' which has been passed or carried; used after the decision has been reached

Seconder: one who supports the 'proposer' of a motion or proposal by 'seconding' it

Secretary: committee official responsible for the internal and external administration of a committee

Secret ballot: a system of voting in secret

Shelve: to drop a motion which has no support

Sine die: from Latin, literally, 'without a day', that is to say indefinitely, eg 'adjourned sine die'

Standing Committee: a committee which has an indefinite term of office

Standing Orders: rules of procedure governing public sector meetings

Table: to introduce a paper or schedule for noting

Taken as read: to save time, it is assumed the members have already read the minutes

Treasurer: committee official responsible for its financial records and transactions

Ultra vires: beyond the authority of the meeting to consider

Unanimous: all being in favour

Personnel

Letters of application

A letter of application for a job is essentially a sales letter. In such a letter you are trying to sell yourself, so your letter must:

- capture attention by using a good writing style
- arouse interest in your qualifications
- carry conviction by your past record and testimonials
- bring about the action you want the prospective employer to take – to grant an interview and eventually give you the job.

Style of application

Unless an advertisement specifies that you must apply in your own handwriting, or the post is purely clerical or bookkeeping, it's always best to send a well-presented typewritten cover letter. A well-displayed, easy-to-read letter will attract attention at once and create a favourable first impression.

Some applicants write a long letter containing lots of information about education, qualifications and experience – I suggest you should avoid this. First of all, when writing like this in your letter it can sound rather boastful, and secondly it makes the information very difficult to find in your letter.

I think it is better to write a short letter applying for the post and stating that your curriculum vitae (or resumé) is enclosed/attached. Then attach a separate document giving all the relevant details about your personal background, education, qualifications and experience. Do not duplicate such information in your covering letter.

 TIP Curriculum vitae is Latin, literally meaning 'the course of one's life'. It sets out all your personal details, education, qualifications and working experience in a logical, easy-to-find column format.

Points of guidance

- Remember the purpose of your application is not to get the job but to get an interview.
- Ensure your application looks attractive and neatly presented; make it stand out from the rest.
- Be brief; give all the relevant information in as few words as possible.
- Write sincerely, in a friendly tone, but without being familiar.
- Do not make exaggerated claims or sound boastful; simply show a proper appreciation of your abilities.
- Do not imply that you are applying for the job because you are bored with your present one.
- Do not enclose originals of your testimonials; send copies with your application but take your originals along to the interview.

A busy employer has little time for long, rambling correspondence. Avoid the temptation to include details in which the recipient is unlikely to be interested, no matter how important they may be to you. You should also avoid generalising, and instead be quite specific in the information you provide. For example, instead of saying 'I have several years of relevant experience in a well-known firm of engineers', state the number of years, state the experience and give the name of the firm.

When you have written your letter, read it carefully and ask yourself these questions:

- Does it read like a good business letter?
- Will the opening paragraph interest the employer enough to prompt him/her to read the rest?
- Does it suggest that you are genuinely interested in the post and the kind of work to be done?
- Is your letter neatly presented and logically structured?

If your answer to these questions is 'Yes', then you may safely send your letter.

Application for an advertised post

Application letter

When your application is in response to an advertisement in a newspaper or journal, always mention this in the opening paragraph or in the subject heading.

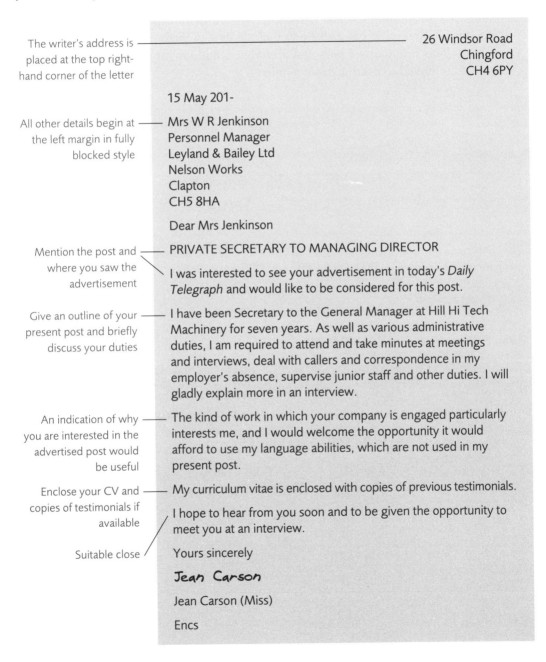

The writer's address is placed at the top right-hand corner of the letter

26 Windsor Road
Chingford
CH4 6PY

15 May 201-

All other details begin at the left margin in fully blocked style

Mrs W R Jenkinson
Personnel Manager
Leyland & Bailey Ltd
Nelson Works
Clapton
CH5 8HA

Dear Mrs Jenkinson

Mention the post and where you saw the advertisement

PRIVATE SECRETARY TO MANAGING DIRECTOR

I was interested to see your advertisement in today's *Daily Telegraph* and would like to be considered for this post.

Give an outline of your present post and briefly discuss your duties

I have been Secretary to the General Manager at Hill Hi Tech Machinery for seven years. As well as various administrative duties, I am required to attend and take minutes at meetings and interviews, deal with callers and correspondence in my employer's absence, supervise junior staff and other duties. I will gladly explain more in an interview.

An indication of why you are interested in the advertised post would be useful

The kind of work in which your company is engaged particularly interests me, and I would welcome the opportunity it would afford to use my language abilities, which are not used in my present post.

Enclose your CV and copies of testimonials if available

My curriculum vitae is enclosed with copies of previous testimonials.

I hope to hear from you soon and to be given the opportunity to meet you at an interview.

Suitable close

Yours sincerely

Jean Carson

Jean Carson (Miss)

Encs

Curriculum vitae

Your curriculum vitae (sometimes called a résumé) should set out all your personal details, together with your education, qualifications and working experience. It should be displayed attractively so that all the information can be seen at a glance. It should not extend to more than two pages. Wherever possible, the information should be categorised under headings and columns. Please note that in this CV the writer has chosen to include date of birth, nationality and marital status. This is actually a personal preference, and law in the UK says employers are not allowed to ask for these details.

CURRICULUM VITAE

Personal details at the beginning →

NAME	Jean Carson
ADDRESS	26 Windsor Road
	Chingford
	Essex CH4 6PY
TELEPHONE	020 8529 3456

It is personal choice as to whether these items are included →

DATE OF BIRTH	26 May 1965
NATIONALITY	British
MARITAL STATUS	Single

State full- and part-time educational courses →

EDUCATION

19– to 19–	Woodford High School
19– to 19–	Bedford Secretarial College (Secretarial Course)

Mention your present job first and work backwards →

WORKING EXPERIENCE

April 201- to present	Personal Secretary to General Manager	Hill Hi Tech Machinery Vicarage Road Leyton London E10 5RG
Sept 19– to March 201-	Shorthand Typist	Bains, Hoyle & Co Solicitors 60 Kingsway London WC2B 6AB

QUALIFICATIONS

List qualifications in full (don't just say '4 A levels')

8 GCE A Levels

7 GCE O Levels

Cambridge	International Diploma in Business Administration	2010
LCCIEB	Executive Secretary's Diploma	2008
LCCIEB	Shorthand – 120 wpm	2006
RSA	140 wpm Shorthand	2006

SPECIAL AWARDS

Mention any special achievements

RSA Silver medal for shorthand 140 wpm

Governors' prize for first place in college examinations

INTERESTS

Hobbies, interests or other relevant information

Music; Languages; Hockey; Golf; Swimming

REFEREES

Give at least two referees – a former employer? a teacher?

1. Dr R G Davies
   Principal
   Bedford Secretarial College
   Righton Road
   Bedford MH2 2BS

2. Ms W Harris
   Partner
   Bains, Hoyle & Co
   60 Kingsway
   London WC2B 6AB

3. Mr W J Godfrey OBE
   Managing Director
   Reliance Cables
   Vicarage Road
   Leyton
   London E10 5RG

Date your CV with month/year

June 201-

Application using an introduction

Sometimes your application will result from an introduction by a friend or colleague. In this case, mention the introduction in the opening paragraph as a useful way of attracting attention.

Dear Mr Barker

Mrs Phyllis Naish, your Training Manager, has told me that you have a vacancy for a Marketing Assistant. I would like to be considered for this post.

As you will see from my enclosed curriculum vitae, I have several A levels as well as secretarial qualifications gained during an intensive one-year course at Walthamstow College of Commerce.

I have been Shorthand Typist in the Marketing Department of Enterprise Cables Ltd for two years and have been very happy there, gaining a lot of valuable experience. However, the office is quite small and I now wish to widen my experience and hopefully improve my prospects.

My former headmistress has written the enclosed testimonial and has agreed to give further details should they be needed. My present employer has also agreed to provide further information.

I am able to attend an interview at any time, and hope to hear from you soon.

Yours sincerely

Application for post of Sales Manager

Dear Sir

Mention the post and advertisement —— I was very interested to see your advertisement for a Sales Manager in yesterday's *Daily Telegraph*, and would like to be considered for this post.

Enclose CV and briefly discuss working experience —— My full particulars are shown on my curriculum vitae, from which you will see that I have had 10 years' experience in the sales departments of two well-known companies. My duties at Oral Plastics Ltd include training sales staff, dealing with the company's

Mention why you are applying —— foreign correspondence, and organising market research and sales promotion programmes. I thoroughly enjoy my work and am very happy here, but feel that my experience in marketing has prepared me for the responsibility of full sales management.

Refer to referees —— Ms Harriet Webb, Sales Manager of my former company, has agreed to provide a reference for me. Her details are shown on my curriculum vitae.

Suitable close —— I shall be pleased to provide any further information you may need, and hope I may be given the opportunity of an interview.

Yours faithfully

Application for a teaching post

This letter of application is sent by a trainee teacher to the Chief Education Officer of her local authority enquiring about suitable teaching posts.

Dear Sir

At the end of the present term I shall complete my one-year teacher training course at Garnett College of Education. I am keen to obtain a post at a school or college in the area administered by your authority.

From my curriculum vitae, you will see that I have 6 O level and 2 A level passes, as well as advanced qualifications in many secretarial subjects. I have held secretarial positions in the London area for a total of 8 years, during which time I studied for my RSA Shorthand and Typewriting Teachers' Diplomas. Having enjoyed the opportunity to teach these subjects in evening classes at the Chingford Evening Institute for 2 years, I was prompted to take up a full-time Certificate in Education at Garnett.

I am looking forward to pursuing a career in teaching, and if there is a suitable vacancy in your area, I hope you will consider me for it.

I look forward to hearing from you and meeting you at an interview.

Yours faithfully

Application for post of Data Processing Trainee

In this letter the writer gives details of his education and qualifications in his letter instead of in a separate curriculum vitae. This style is useful when the applicant does not have enough previous working experience to warrant a CV.

Dear Sir

I would like to apply for the post of Management Trainee in your Data Processing Department advertised today in *The Guardian*.

I obtained A level passes in Mathematics, Physics and German at Marlborough College, Wiltshire. The College awarded me an open scholarship to Queens College, Cambridge, where I obtained a first in Mathematics and a second in Physics. After leaving university last year I accepted a temporary post with Firma Hollander & Schmidt in order to improve my German and gain some practical experience in their laboratories at Bremen. This work comes to an end in 6 weeks' time.

My special interest for many years has been computer work and I should like to make it my career. I believe my qualifications in Mathematics and Physics would enable me to do so successfully.

I am single and would be willing to undertake the training courses away from home to which you refer in your advertisement.

My former Housemaster at Marlborough, Mr T Gartside, has consented to act as my referee (telephone 0117 234575), as has Dr W White, Dean of Queens College, Cambridge (telephone 01246 453453). I hope that you will take up these references and grant me the opportunity of an interview.

Yours faithfully

An unsolicited application

An unsolicited application (ie uninvited) is the most difficult to write since there is no advertisement or introduction to tell you anything about the work, or indeed whether there is a vacancy. In such a situation you must try to find out something about the company's activities and then show how your qualifications and experience could be used.

Dear Sir

I believe that a large and well-known organisation such as yours might be able to use my services. For the past 8 years I have been a Statistician in the Research Unit of Baron & Smallwood Ltd, Glasgow. I am now looking for a change of employment that would widen my experience and at the same time improve my prospects.

At the University of London I specialised in merchandising and advertising, and was awarded a PhD degree for my thesis on 'Statistical Investigation in Research'. I thoroughly enjoy working on investigations, particularly where the work involves statistics.

Although I have had no experience in consumer research, I am familiar with the methods employed, and fully understand their importance in the recording of buying habits and trends. I would love to use my services in this type of research and hope you will invite me to attend an interview. I could then give you further information and bring testimonials.

I hope to hear from you soon.

Yours faithfully

 TIP When writing to someone for the first time, remember: there is no second chance to create a great first impression!

Testimonials

A testimonial is a written recommendation of a person's character and ability written by an employer when an employee leaves their employment. An employee can then keep all received testimonials together to be shown to future potential employers.

Testimonials may be given to an employee leaving your company if requested, providing that this is in accordance with company policy. They are voluntary, but if one is written it is useful to include:

- Start and finish dates of employment
- Post(s) held
- Details of the duties carried out
- Work attitude and personal qualities
- A recommendation

If you do mention anything to do with the employee's work attitude or personal qualities, it must be factually accurate and not present an overly negative or positive message.

 TIP Some of the wording in testimonials may be useful for references, but do remember that references are more functional whereas testimonials may be more personal.

Formal testimonial for Secretary

This testimonial was requested by an employee who worked at a company for a period of 8 years until she took up teacher training.

TO WHOM IT MAY CONCERN

Miss Sharon Tan was employed as Shorthand Typist in this company's Sales Department when she left secretarial college in July 19–. She was promoted to my Personal Secretary in 201- until she left the company in March 201-.

Duration of employment/position

Her responsibilities included the usual secretarial duties involved in such a post as well as attending meetings, transcribing minutes and supervising and advising junior secretaries.

Duties

Sharon used her best endeavours at all times to perform her work conscientiously and expeditiously. She was an excellent secretary, an extremely quick and accurate shorthand typist, and meticulous in the layout, presentation and accuracy of her work. I cannot overstress her exceptional work rate which did not in any way detract from the very high standards she set for herself.

Working attitude

Sharon had a great working attitude, and she made a point of building great team relationships. It was a great loss to both myself and the company when Sharon took up teacher training.

Personal qualities

In my opinion, Sharon has the necessary character, dedication and approach to be suitable for the position of personal secretary or to enter the teaching profession. I can recommend her highly and may be contacted for further information.

Recommendation

Ian Henley
Deputy Chairman

Testimonial for Head of Department

Here is another very favourable testimonial that was issued to someone who left a private college after completing a two-year contract as Head of Department.

TO WHOM IT MAY CONCERN

Norman Tyler has been employed by this College as Head of Business Studies from August 201- to 9 March 201-.

As well as capably handling the responsibilities for the overall administration of his department, Norman ably taught Economics, Commerce and Management Appreciation to students of a wide range of ability and age groups on courses leading to Advanced LCCI examinations.

Norman is a highly competent and professional teacher whose class preparation is always thorough and meticulous. His committed approach to teaching is matched by his administrative abilities. He has made a substantial contribution to course planning, student counselling, curriculum development and programme marketing.

Norman possesses an outgoing personality and he mixes well. He makes his full contribution to a team and is popular with his students and colleagues alike.

I am confident that Norman will prove to be a valuable asset to any organisation fortunate enough to employ him. It is with pleasure that I recommend him highly and without hesitation.

Faisal Shamlan
Principal

 CAUTION Do check the laws and legislation in your own country about what may or may not be included in a testimonial or a reference.

Testimonial for colleague

April 201-

TO WHOM IT MAY CONCERN

State how you know the person — I have known Sonja Bergenstein for several months in her capacity as Business Development Executive of SingaJobs.com, Singapore.

As a freelance Training Consultant, I have worked with SingaJobs.com on many occasions. They have acted as my agent in marketing and promoting my two-day workshops 'Transform your Business Writing Skills', and Sonja has been one of the team working on this.

My contact with Sonja has been mainly on the days of the workshops, when she has always been well-organised, helpful and friendly. She has a very sociable and pleasant personality, and she always goes the extra mile in making sure that all participants are kept happy during each workshop. She has been an expert in *Mention some key points about work/ responsibilities* — public relations during breaks, when it is important to mix with participants and make sure they are well looked after. When it has been necessary to address groups, Sonja has always been confident and able to express herself clearly and with interest.

Give additional comments on personality and attitude — Working with many different nationalities in the Singaporean scene has been no problem for Sonja. She had adapted well to the different cultures, and she has been able to mix and get on well with people from all races. From my experience, she has been very well liked and respected, a hard-working member of the team at SingaJobs.com who will be sorely missed.

Close with a recommendation — I have certainly enjoyed working with Sonja, and wish her every success in her future career, in which I am sure she will do extremely well.

Shirley Taylor
Author and Training Consultant

 CAUTION It's always best to check the current laws and legislation in your country to make sure you are following official guidelines when dealing with any personnel issue.

References

Even if testimonials are provided at the time of sending an application letter, it is usual to state (either on your CV or covering letter) the names of one or two people who have consented to act as referees. Prospective employers may contact such referees by either telephone or letter to obtain further information about an applicant's work performance.

When supplying a reference for a previous employee, you must act in accordance with company policy and in line with previous requests. Some companies choose only to provide confirmation that the employee worked there previously, but others may provide more information as long as it is factually accurate. If a reference is provided, it is useful to include:

- Start and finish dates of employment
- Post(s) held

If you do mention anything to do with the employee's work attitude or personal qualities, it must be factually accurate and not present an overly negative or positive message.

 CAUTION Do check the laws and legislation in your own country about what may or may not be included in a testimonial or a reference.

Letter taking up a reference

Dear Mrs Lambert

Mention applicant's name and post applied for — Mr James Harvey, who I believe is your Foreign Correspondent, has applied to us for a similar post and has given your name as a referee.

Ask for information about his work — Please can you let me know if his services with you have been entirely satisfactory and if you consider he would be able to accept full responsibility for the French and German correspondence in a large and busy department.

Include specific details regarding ability — I am aware that Mr Harvey speaks fluent French and German, but I am particularly interested in his ability to produce accurate translations into these languages of letters that may be dictated to him in English.

Give an assurance of confidentiality — Thank you in advance for any information you can provide, in strict confidence of course.

Yours sincerely

Favourable reply

In this reply, the writer recommends the employee very highly and without hesitation, feeling confident that he can carry out the duties in the post.

Dear Mr Brodie

I am pleased to be able to reply favourably to your enquiry of 6 April concerning Mr James Harvey.

Mr Harvey is an excellent linguist and for the past 5 years has been in sole charge of our foreign correspondence, most of which is with European companies, especially in France and Germany.

We have been extremely pleased with the services provided by Mr Harvey. If you engage him, you may rely upon him to produce well-written and accurate transcripts of letters into French and German. He is a very reliable and steady worker and has an excellent character.

We wish him success, but at the same time shall be very sorry to lose him.

Yours sincerely

Cautious reply

In this reply the writer is very cautious, implying that the applicant lacks the experience needed for control of a department. However, the writer is very careful not to come straight out and say this in so many words.

Dear Mr Brodie

Thank you for your letter of 6 April concerning Mr James Harvey.

Mr Harvey is a competent linguist, and for the past 5 years has been employed as senior assistant in our foreign correspondence section. He has always been conscientious and hard-working. It's difficult for me to judge if he would be capable of taking full responsibility for a large and busy department, as his work here has always been carried out under supervision.

If you require any further information please do not hesitate to contact me.

Yours sincerely

Enquiry letter requesting a reference

In this letter, another prospective employer requests information about the work and character of an applicant.

Dear Mr Jones

Mr Lionel Picton has applied to us for an appointment as Manager of our factory in Nairobi. We are leading manufacturers of engineered components used in the petrochemical industry and are looking for a qualified engineer with works manager's experience in medium or large batch production.

Mr Picton tells us that he is employed by you as Assistant Manager of your factory in Sheffield. We would appreciate your thoughts on his competence, reliability and general character.

The information you provide will be treated in strictest confidence.

Yours sincerely

Favourable reply

Dear Mr Gandah

Acknowledge letter and give background information — Thank you for your letter of 6 August regarding Mr Lionel Picton, who has been employed by this company for the past 10 years.

Give details about the applicant's work, qualifications and attitude — Mr Picton served his apprenticeship with Vickers Tools Ltd in Manchester, followed by a part-time course for the Engineering and Work Study Diploma of the Institution of Production Engineers. He is technically well-qualified, and for the past 5 years has been our Assistant Works Manager responsible for production and associated activities in our Sheffield factory. In all aspects of his work he has shown himself to be hard-working, conscientious and in every way a very dependable employee.

Finish with a recommendation and personal word about the applicant — I can recommend Mr Picton without the slightest hesitation. I feel sure that if he was appointed to manage your factory in Nairobi he would bring to his work a genuine spirit of service, which would be found stimulating and helpful by all who worked with him.

Yours sincerely

Applicant's thank you letter

Those who have provided references will naturally be pleased to know if the applicant was successful or not. It's always a good idea to inform and thank those who supported them.

Dear Mr Freeman

Many thanks for supporting my application for the post as Manager of the Barker Petrochemical Company in Nairobi.

I know that the generous terms in which you wrote about me had much to do with my being offered the post, and I am very grateful to you for the reference you provided for me.

I have always appreciated your help and encouragement. Thank you again.

Yours sincerely

Enquiry using numbered points

In this enquiry the writer is looking for certain qualities. To make sure that each one is covered in a reply, numbered points are used.

Dear Miss French

Introduction states name of applicant and post applied for —— Miss Jean Parker has applied for a post as Administrator in our Sales Department. She states that she is presently employed by you and has given your name as a referee.

I hope you can answer these questions regarding her abilities and character:

Specific questions regarding the applicant are numbered and listed ——
1. Is she conscientious, intelligent and trustworthy?
2. Is she capable of dealing with any difficult situations?
3. Are her keyboarding and administrative skills satisfactory?
4. Is she capable of dealing accurately with figure work?
5. Is her output satisfactory?
6. Does she get on well with her colleagues?

Give an assurance of confidentiality —— We will treat any information you provide in strict confidence.

Yours sincerely

Reply

Dear Mr Kingston

Thank you for your letter of 15 April.

Miss Jean Parker has been employed as Assistant Sales Administrator in our general office for the past two years, and I feel sure that you will find her in every way satisfactory. I have nothing but good to say about her.

In reply to each of the questions in your letter, I have no hesitation in saying that Miss Parker meets all these requirements.

We will be sorry to lose Miss Parker, but realise that her abilities demand wider scope than is possible at this company.

Yours sincerely

Favourable reference – former student

Dear Mrs Thompson

MISS CAROLINE BRADLEY

Thank you for your letter dated 3 June. I welcome the opportunity to support Miss Bradley's application for the post of your Marketing Assistant.

Miss Bradley was a student at this college during the year 201- to 201-. Admission to this intensive one-year course is restricted to students with good school-leaving qualifications. The fact that Miss Bradley was admitted to the course is in itself evidence of excellent academic ability. Upon completing her course, she was awarded the title 'Student of the Year', being the student gaining highest qualifications over the one-year course.

In all other respects, Miss Bradley's work and attitude were entirely satisfactory, and I can recommend her to you with every confidence. I feel sure that if she was appointed she would perform her duties diligently and reliably.

Yours sincerely

Favourable reference – Department Manager

Dear Mr Lee

Mr Leonard Burns is both capable and reliable. He came to us five years ago to take charge of our Hardware Department.

Leonard knows the trade thoroughly and does all the buying for his department with great success. I know that for some time he has been looking for a similar post with a larger store. While we would be sorry to lose his services, we would not wish to stand in the way of the advancement that could be offered by a store such as yours.

Please contact me again if you have any questions.

Yours sincerely

Useful sentences for references and testimonials

- Mr John Smith was employed by this company from ___ to ___.
- I am very glad of this opportunity to support Miss Lim's application for a position in your company.
- Mr Johnson has proven to be an efficient, hard-working, trustworthy and very personable employee.
- Sharon used her best endeavours at all times to perform her work conscientiously.
- In my opinion, Harrison has the necessary character, dedication and approach to be suitable for the position of personal secretary.
- Nigel has an outgoing personality and he mixes well.
- He makes his full contribution to a team and is popular with his colleagues and clients.
- Geetha made a substantial contribution to the work of the Sales Department and always performed her work in a businesslike and reliable manner.
- It was a great loss to both myself and the company when Miss Turner moved abroad.
- We were very sorry to lose Miss Fisher and she will be greatly missed.
- It is with great pleasure that I recommend Martha Tan for this position in your company.
- I can recommend Mr Cheong without hesitation, and know you will find him an excellent addition to your staff.

■ We were very sorry to lose Miss Franks and are pleased to recommend her highly and without hesitation.

Interview letters

If a lot of applications are received for a post, it is unlikely that all applicants can be interviewed. In such cases a shortlist will be drawn up of those applicants thought to be most suitable for interview. Letters should also be sent to the unsuccessful applicants.

Invitation to attend for interview

A letter inviting an applicant for interview should first acknowledge receipt of the application, and then go on to give a day, date and time for the interview. Confirmation is often requested.

Dear Miss Wildman

SENIOR SECRETARY TO TRAINING MANAGER

Thank you for your application for this post.

You are invited to attend for an interview with me and Mrs Angela Howard, Training Manager, on Friday 29 May at 3.30 pm. Please come to our main reception and you will then be shown to the second floor. Our offices have special access lifts if required.

Please let me know either by letter or telephone whether this appointment will be convenient for you or if any reasonable adjustments need to be made.

Yours sincerely

Confirmation of attendance

Dear Mrs Graham

SENIOR SECRETARY TO TRAINING MANAGER

Thank you for your letter inviting me to attend for interview on Friday 29 May at 3.30 pm.

I shall be pleased to attend and look forward to meeting you and Mrs Howard.

Yours sincerely

Letter of rejection before interview

It is courteous to write to the applicants who have not been included on the shortlist. The letter can be worded in such a way that it does not cause offence or negative feelings.

Dear

Thank you for your application for the post of Senior Secretary to the Training Manager.

We have received many applications for this post. I am afraid that your experience and qualifications do not match all our requirements closely enough, so we cannot include you on our shortlist for this post.

Thank you for the considerable time and effort you put into preparing your application. You have a lot of useful experience, and I am sure that you will soon find suitable employment.

Yours sincerely

Job description

Job description for Senior Secretary

A job description gives details of the duties and responsibilities involved in a post, including any supervisory duties, specific authority and any special features of the post.

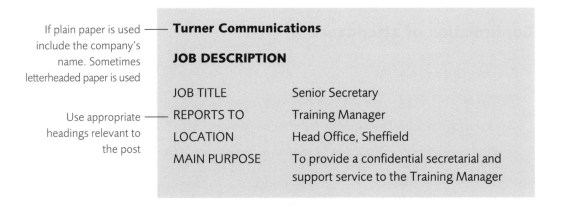

If plain paper is used include the company's name. Sometimes letterheaded paper is used

Turner Communications

JOB DESCRIPTION

Use appropriate headings relevant to the post

JOB TITLE	Senior Secretary
REPORTS TO	Training Manager
LOCATION	Head Office, Sheffield
MAIN PURPOSE	To provide a confidential secretarial and support service to the Training Manager

Sometimes specific —— REQUIREMENTS
requirements of the post
holder are included

1. Abilities: use initiative, decide priorities, work without supervision
2. Previous experience at senior level
3. Skills: Microsoft Office 10, note-taking skills, good organiser, good interpersonal skills, excellent writing skills
4. High standard of education with appropriate secretarial/administration qualifications

List the main duties —— MAIN DUTIES AND RESPONSIBILITIES
and responsibilities

1. Provide secretarial support to the Training Manager.
2. Deal with mail, answer telephone enquiries, take messages and compose correspondence.
3. Take shorthand dictation and deal with instructions from manuscript, audio or disk and to transcribe documents accurately and consistently.
4. Maintain the diary of the Training Manager.
5. Arrange meetings and produce accurate minutes.
6. Arrange training courses and seminars.
7. Make travel and accommodation arrangements as may be required.
8. Ensure the security of the office and confidential documents.

Finish with this —— 9. Carry out any other duties as may be expected in a post of this
standard clause   level.

ST/BT

June 201-

TIP Note how all the points in this job description begin with verbs. This is called grammatical parallelism, or grammatical construct. You can read more about this in Chapter 16.

Job description for Telephone Executive

JOB DESCRIPTION

Job Title	Telephone Executive (Marketing)
Location	Marketing Department, Head Office
Responsible to	Marketing Manager
Main Purpose of Job	To telephone customers with the objective of identifying opportunities where business can be increased

MAIN DUTIES AND RESPONSIBILITIES

1. Achieve daily call rate targets and any target set for sales campaigns.
2. Be courteous to customers by using a good telephone manner at all times.
3. Carry out any administrative requirements generated by the telephone calls in an accurate and efficient manner. This may include sending letters, fax messages, emails, reports, product literature, etc.
4. Undertake training courses to make good use of telephone selling techniques.
5. Undertake training on the company's products and services and promote associated products where appropriate.
6. Carry out competitor market research by contracting their branches to gather information on pricing, product availability, etc, as directed by your supervisor.
7. Carry out any other tasks as requested by your supervisor.

Offers of appointment

Letters appointing staff should state clearly the salary and any other conditions of appointment. If the duties of the post are described in detail on a Job Description and enclosed with the letter, it will not be necessary to duplicate such details in the letter itself.

Letter confirming offer of employment

If an appointment is made verbally at the interview, it should be confirmed by letter immediately afterwards.

Dear Miss Wildman

Offer the job and include a commencement date —— I am pleased to confirm the offer we made to you yesterday of the post of Senior Secretary to the Training Manager, commencing on 1 August 201-.

Specify the duties or enclose Job Description and Contract of Employment —— Your duties will be as we discussed at the interview and as described on the attached Job Description. Our official Contract of Employment is also enclosed.

Include details of salary and holidays —— This appointment carries a commencing salary of £18,000 per annum, rising to £19,500 after one year's service and thereafter by annual review. You will be entitled to 4 weeks' annual holiday.

Mention termination information —— The appointment may be terminated at any time by either side giving two months' notice in writing.

Ask for confirmation —— Please confirm that you accept this appointment on the terms stated and that you will be able to commence your duties on 1 August.

Yours sincerely

Letter offering appointment

When the appointment is not made at the interview, the offer will be made by letter to the selected applicant as soon as possible.

Dear Miss Jennings

Thank you for attending the interview yesterday. I am pleased to offer you the post of Secretary in our Sales Department at a starting salary of S$1,500 (Singapore dollars) per month. Your commencement date will be Monday 1 October 201-.

As discussed, office hours are 0900 to 1730 with one hour for lunch. Your holiday entitlement will be as shown on the attached Contract of Employment.

Please return a signed copy of this letter of employment confirming your acceptance of this appointment on these terms.

I look forward to hearing from you soon.

Yours sincerely

Acceptance of offer of employment

Any offer letter should be accepted in writing immediately. Many employers send two copies of an offer letter and ask for the second copy to be signed and returned.

Dear Miss Tan

Thank you for your letter of 24 August offering me the post of Secretary in your Sales Department.

I am pleased to accept this post on the terms stated in your letter, and confirm that I can commence work on 1 October. My signed copy of your offer of employment is attached.

I am looking forward to joining your organisation.

Yours sincerely

Declining an offer of employment

If you do not wish to take up the offer of employment it is courtesy either to call the person to let them know you don't wish to accept, or to send a brief email. In this way the employer may make a second choice as soon as possible.

Dear Miss Tan

Thank you for your letter of 24 August offering me the post of Administrative Executive in your company.

I am sorry that I will be unable to take up this position. My present employer has discussed with me the company's plans for expansion and I have been offered the new post of Office Manager. This will offer me a challenge that I feel I must accept.

Thank you for your understanding.

Yours sincerely

Letter to unsuitable applicants

As soon as an offer of employment has been accepted by the selected applicant, it is courteous to write letters to the remaining applicants who were interviewed telling them that their application was unsuccessful.

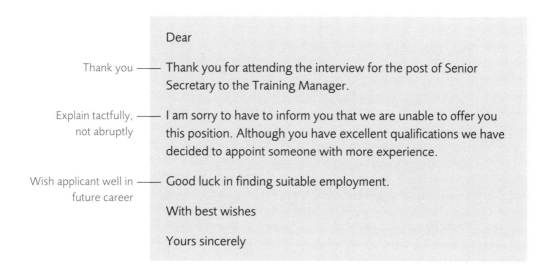

Dear

Thank you ——— Thank you for attending the interview for the post of Senior Secretary to the Training Manager.

Explain tactfully, ——— I am sorry to have to inform you that we are unable to offer you
not abruptly this position. Although you have excellent qualifications we have decided to appoint someone with more experience.

Wish applicant well in ——— Good luck in finding suitable employment.
future career

With best wishes

Yours sincerely

Termination of employment

By the Employment Rights Act 1996, employees who feel they have been unfairly dismissed have the right to appeal to an Employment Tribunal. An employer must be able to show that the dismissal was justified by referring to the employee's conduct or inability/failure to do the job satisfactorily.

Where it is decided to terminate the employment of a person whose services have been unsatisfactory, the company should always follow their policies and procedures and ensure they are following statutory guidelines. A lot of help is available online, together with standard templates, so it is not felt appropriate for us to include guidelines or sample documents in this chapter.

Employee's letter of resignation

A Contract of Employment made for a stated period usually comes to an end when the period is completed unless both parties agree to an extension. If the contract is for an unstated period, there may be a clause in it that the employment may be ended at any time by either of the parties giving an agreed period of notice. Again, it is always best to check practices in your country.

Dear Miss Ward

I am sorry to inform you that I wish to give two weeks' notice of my resignation from the company. My last day of work will be 30 June 201-.

I have been very happy working here for the past two years and found my work challenging and enjoyable. However, I have obtained a post in which I will have more responsibilities and greater career prospects.

Thank you for your help and guidance during my employment.

Yours sincerely

Sundry personnel matters

Transfer of employee to other work

Where it is necessary to transfer an employee from work that the employee has enjoyed, you must clearly explain the reasons for the transfer and any advantages or benefits. Perhaps there will be the prospect of more interesting and responsible work, more experience, better pay, improved prospects. With tact, it should be possible to convey what may be unwelcome or disappointing news to an employee without causing hurt feelings or offence. In this way, what might otherwise be received as unwelcome news may almost be turned into good news.

In this case, a long-standing employee is happily settled into a routine with no wish to change, but this has been made necessary due to technological developments within the company.

Dear Mr Turner

As Mrs Williamson has already discussed with you, we have arranged to appoint you as Section Supervisor in the Stores Department with effect from Monday 1 July. Your salary will be £19,200 per annum.

In your new post you will report directly to Mr James Freeman, Storekeeper, and you will be responsible for the work of the clerical staff employed in the department.

The management has greatly appreciated your 30 years of loyal service in the Invoice Department, and we are sorry that it is necessary to move you from a department with which you are so familiar. Our only reason for doing so is that invoicing will be completely changed by the introduction of computerised methods. We feel sure that you will understand that it is uneconomic for us to retrain our long-standing employees who might find difficulty in adjusting to new ways of working.

In your new post you will find ample scope for your experience. I know you will do a good job and hope you will find it enjoyable.

Yours sincerely

Recruitment of staff through an agency

Employers in need of new office staff frequently make their requirements known to employment agencies. Such agencies will introduce either full-time, part-time or temporary staff in return for a commission related to the amount of wage or salary paid.

Dear Sir/Madam

I hope you will be able to help me to fill a vacancy that has just arisen in my department.

My Secretary needs administrative help on a part-time basis. This will be an interesting post and ideal for someone who wishes to work for only a few hours each week. Applicants of any age would be considered, but willingness and reliability, plus attention to detail, are preferable to someone with high qualifications.

Ideally, the successful applicant will be required to work for three hours each morning from Monday to Friday. We would be willing to consider an alternative arrangement if necessary.

I propose payment based on an hourly rate of £5 to £6 according to age and experience.

Please let me know whether you have anyone on your register who would be suitable.

Yours faithfully

Request for an increase in salary

Any letter requesting an increase in salary should be worded very carefully, explaining tactfully the reason why you feel a salary increase is justified.

Dear Mr Browning

My present appointment carries an annual salary of £18,500; this was reviewed in March last year.

During my five years with the company, I feel I have carried out my duties conscientiously and have recently acquired additional responsibilities.

I feel that my qualifications and the nature of my work justify a higher salary and I have already been offered a similar position with another company at a salary of £20,000 per annum.

My present duties are interesting and I thoroughly enjoy my work. Although I have no wish to leave the company, I cannot afford to turn down this offer unless some improvement in my salary can be arranged.

I hope a salary increase will be possible, and look forward to hearing from you soon.

Yours sincerely

Letter of resignation

When you decide to leave a company you must hand in your notice. It is usual to do so with a formal letter of resignation in accordance with the company's conditions of employment.

Dear Mr McKewan

Please accept notice of my intention to leave the company in one month's time, ie 28 July.

As I have discussed with you I have accepted a position with another company that will allow me greater responsibilities and improved opportunities for advancement.

Thank you for your support during my two years with Turner Communications. I have gained a lot of valuable experience and am sure this will be very useful.

Yours sincerely

Welcome letter to new employee

I am pleased that you have chosen to accept our offer of employment as Business Development Executive, and hope that this will be the beginning of a mutually beneficial association.

We encourage our staff to take advantage of selected courses that are available in Singapore, so as to improve their skills and learn new skills in related areas. The current list of courses and their corresponding registration dates are posted on the employee bulletin board. If you decide to attend one of these courses, please ask your supervisor to make the necessary arrangements.

If you have any questions at any time, please give me a call on [telephone number].

Once again, welcome to SingComm Pte Ltd.

Reminder about health and safety policy

In view of the recent unfortunate accident involving a visitor to our premises, I would like to remind you about our health and safety policy.

Whenever you have a visitor to the building, you are responsible for their health and safety at all times. For his/her own safety, a visitor should not be allowed to wander freely around the building. If, for example, a visitor needs to use the washroom, you should accompany them and escort them back.

For security reasons, if you see someone you do not recognise wandering around the building unaccompanied, do not be afraid to ask questions. Ask politely why he or she is here, which member of staff they are visiting, and if that person knows the visitor has arrived.

All SingComm employees have a responsibility to take reasonable care of themselves and others and to ensure a healthy and safe workplace. If you notice any hazard or potential hazard, please bring it to the attention of the Health and Safety Manager, Michael Wilson, who will investigate the issue.

A copy of the company's health and safety policy is attached. Please take a few minutes to read it through to remind yourself of the main points.

Thank you for your help in ensuring the health and safety of all employees and visitors to SingComm.

Letter regarding outstanding holiday entitlement

In the past it has been a policy of the company that all staff must take their holiday entitlement within one calendar year. Any holiday entitlement not taken before 31 December each year has been forfeited.

I am now glad to announce that this rule has been amended to provide staff more flexibility with holidays.

With immediate effect anyone who has up to five days' holiday entitlement outstanding at 31 December may carry this over to 31 March the following year. Any days that have not been used by 31 March will be forfeited. Unused holiday entitlement may not be converted to pay in lieu.

Your manager/supervisor will still be required to approve staff leave. This will take into account the business and operational needs of the department and especially clashes with other staff.

If you have any questions about this new policy, please telephone the Human Resource Department on extension 456.

Useful expressions

Application letters

Openings

- I wish to apply for the post ___ advertised in ___ on ___.
- I was interested to see your advertisement in ___ and wish to apply for this post.
- I am writing to enquire whether you have a suitable vacancy for me in your organisation.
- I understand from Mr ___, one of your suppliers, that there is an opening in your company for ___.
- Mrs ___ informs me that she will be leaving your company on ___. If her position has not been filled, I would like to be considered.

Closes

- I look forward to hearing from you and to being given the opportunity of an interview.
- I hope you will consider my application favourably.
- I look forward to the opportunity of attending an interview when I can provide further details.

References

Openings

- Mr ___ has applied to us for the position of ___. We hope you can give us your opinion of his character and abilities.
- We have received an application from Miss ___ who has given your name as a referee.
- I am very glad of this opportunity to speak in support of Miss ___'s application for a position in your company.
- Ms ___ has been employed as ___ for the past two years.

Closes

- I will appreciate any information you can provide.
- We will treat all information provided in strictest confidence.
- I am sure you will be more than satisfied with the work of Mr ___.
- I shall be sorry to lose ___ but realise that her abilities demand wider scope than is possible at this company.

Offers of employment

Openings

- Thank you for attending the interview on ___. I am pleased to offer you the position of ___.
- I am pleased to confirm the offer we made to you when you came for interview on ___.
- I was good to meet you last week. I am pleased to offer you the position of ___ commencing on ___.

Closes

- Please let me have written confirmation of your acceptance of this post soon.
- Please confirm in writing that you accept this appointment on the terms stated.
- We look forward to welcoming you to our staff and hope you will be very happy in your work here.

Termination of employment

Openings

- I wish to terminate my services with this company with effect from ___.
- I am writing to confirm that I wish to tender my resignation. My last date of employment will be ___.
- As my family has decided to emigrate, I am sorry to have to tender my resignation.

Closes

- I have been very happy working here and am grateful for your guidance during my employment.
- I am sorry that these circumstances make it necessary for me to leave the company.
- We have been extremely satisfied with your services and hope that you will soon find another suitable post.
- I hope you will soon find alternative employment, and wish you all the best for the future.

Testimonials

Openings

- Mr ___ has been employed by this company from ___ to ___.
- Miss ___ worked for this company from leaving college in 201- until she emigrated to Canada in March 201-.

Central section

- Miss ___ enjoys good health and is a good time-keeper.
- She uses her best efforts at all times to perform her work expeditiously.
- She has always been a hard-working and conscientious employee.
- Miss ___ made a substantial contribution to the work of the ___ department, and always performed her work in a businesslike and reliable manner.
- Mr ___ gave considerable help to his colleagues in improving teaching methods and materials. He also produced many booklets of guidance that are very valuable to other teachers.

Closes

- I have pleasure in recommending ___ highly and without hesitation.
- We hope that ___ meets with the success we feel he deserves.
- I shall be sorry to lose John, but realise that his abilities demand wider scope than are possible at this company.
- I can recommend Miss ___ to you with every confidence.

Reports and proposals

Introduction to reports and proposals

Many different types of report are used in business – some quite short and informal, others fairly lengthy and formal. The ultimate purpose of any report is to provide the foundation for decisions to be made and action taken.

Some reports contain no more than a simple statement recording an event, a visit or some circumstances, with a note of action taken. Other reports include detailed explanations of facts, conclusions, and perhaps recommendations for action.

More detailed reports require a lot of research. This may involve interviews, visits, questionnaires and investigations. The information may be presented in written, tabular or graphic form, and the writer needs to produce clear conclusions and recommendations.

The skills in writing a proposal are the same as in writing a report. However, there are certain differences between these two documents:

Reports	Proposals
■ contain information about what has happened in the past	■ examine what may happen in the future
■ aim mainly to provide information	■ aim mainly to persuade the reader to make a specific decision
■ record objective facts	■ express opinions, albeit supported by objective facts

TIP A well-presented title page on your report will create a good impression.

The Plain English Guide To Writing Reports

I would like to thank the Plain English Campaign for granting me permission to reproduce its *Plain English Guide to Writing Reports* in Appendix 3. This is an excellent guide going through all the principles of report writing – in plain English of course. To find out more about the Plain English Campaign please visit **www.plainenglish.co.uk**.

Memorandum report

MEMORANDUM

To:	Jean Lee, Manager
From:	Sally Turner, Administration Assistant
Ref:	JL/ST
Date:	20 April 201–

VISIT OF MR HO CHWEE LEONG, WANCHAI IMPORTING COMPANY, HONG KONG

Mr Ho Chwee Leong is to visit us on 28 May. As we can expect a large order from his company, Wanchai Importing Company of Hong Kong, it is important that he receives a good impression of our company. Here are the arrangements for the visit.

ARRANGEMENTS MADE

1. Accommodation has been arranged for Mr Ho at Hotel Moderne and I have arranged for a taxi to collect him from the hotel at 9.30 am to bring him to Shazini Shoes factory.
2. When Mr Ho arrives at Shazini Shoes at 10.00 am he will be met by Mr Lee and senior staff who will take him on a visit of the factory.
3. A buffet lunch has been arranged in the guest room at 12.30 pm. Vegetarian food has been arranged.
4. The boardroom has been booked for a conference for the whole afternoon for Mr Ho, Mr Lee and senior staff. Refreshments have been laid on during the afternoon.
5. A taxi has been booked to take Mr Lee to the airport at 5.30 pm, so he can check in for his flight before 6.00 pm.

ARRANGEMENTS STILL TO BE MADE

1. Up-to-date price lists, catalogues and samples of shoes will be provided in the boardroom.
2. Staff will be informed that Mr Lee and senior staff will not be available next Friday.

 TIP Don't use too many different fonts or sizes in your report, as it will not be attractive and could confuse the reader.

Stand-alone report

**MARUMAN STORES, NOTTING HILL BRANCH
REPORT ON POSSIBILITY OF OPENING A CRÈCHE**

INTRODUCTION

State who asked for the report and what you were asked to do — I was asked to investigate the opening of a crèche at the Notting Hill branch by Mrs Lillian Cheng. In order to do this the following steps were taken.

List the steps taken to gather the information —
1. I obtained a breakdown of figures showing the number of customers with young children.
2. I discussed this issue with several customers who brought children to the store.
3. The accommodation, staffing and insurance issues were considered.
4. I investigated the experience of other shops that already have a crèche.

State the findings in a logical order. Use reported speech — FINDINGS

1. 7.3% of Maruman customers have at least one child under the age of 3.
2. The majority of customers interviewed said they would use a crèche if the cost was reasonable. Some of these customers also commented that other friends who are not presently customers might also consider using the shop if there was a crèche.

3. There are strict laws and regulations concerning accommodation and staffing of a crèche. The site would have to be approved to run a crèche before we could start one.
4. Staff appointed to run the crèche would have to be fully qualified.
5. A suitable space would have to be found. This would require running water as well as toilets. The crèche would have to be close to the store entrance but due to noise levels it should be kept separate from the main store.
6. The company would be required to ensure adequate insurance.
7. Many rival stores in the neighbourhood are offering crèche facilities.

CONCLUSIONS

Make a logical conclusion from the findings

A crèche would be popular and well-used if we decided to go ahead with this.

RECOMMENDATIONS

Suggest action that should be taken

I suggest that the company should give further consideration to offering a crèche and investigate the financial aspects that would be involved.

Sign the document and state your name and title

Sally Turner

Customer Services Executive

Include a reference and date

LC/ST
20 April 201-

TIP Make sure all the information in your report is well-researched and substantiated.

CAUTION Always consider carefully the format and section headings of your report. You must make sure they are relevant to the content as well as the intended audience.

A longer proposal

This proposal is reproduced with permission from *How to Write Proposals and Reports that get Results* written by Ros Jay and published by Prentice Hall.

FLEXIBLE WORKING HOURS
An initial study for ABC Ltd
by
Jane Smith

Objective
To identify the factors involved in introducing flexible working hours, to examine their benefits and disadvantages and to recommend the best approach to take.

Summary
At present, almost all employees of ABC Ltd work from 9.00 to 5.00. A handful work from 9.30 to 5.30.

Many, though not all, staff are unhappy with this and would prefer a more flexible arrangement. Some are working mothers and would like to be able to take their children to and from school. Some, particularly the older employees, have sick or elderly relatives who make demands on their time which do not fit comfortably with their working hours.

For the company itself, this dissatisfaction among staff leads to low morale and reduced productivity. It also makes it harder to attract and retain good staff.

There are three basic options for the future:

1. *Leave things as they are*. This is obviously less demanding on resources than implementing a new system. At least we know it works even if it isn't perfect.
2. *Highly flexible system*. Employees would clock on and clock off anytime with a 12$\frac{1}{2}$ hour working day until they have 'clocked up' 35 hours a week. This would be the hardest system to implement.
3. *Limited flexibility*. Staff could start work any time between 8.00 to 10.00 am and work through for eight hours. This would not solve all employees' problems but it would solve most of them.

Proposal
Introduce a system of limited flexibility for now, retaining the option of increasing flexibility later if this seems appropriate.

▶

◄

Position

The current working hours at ABC Ltd are 9.00 to 5.00 for most employees, with a few working from 9.30 to 5.30.

Problem

Although this works up to a point, it does have certain disadvantages, both for the organisation and for some of the employees.

The organisation: The chief disadvantage of the current system is that many of the staff are dissatisfied with it. This has become such a serious problem that it is becoming harder to attract and retain good staff. Those staff who do join the company and stay with it feel less motivated: this, as research has shown, means they are less productive than they could be.

The employees: Some employees are satisfied with their current working hours, but many of them find the present system restrictive. There are several reasons for this but the employees most strongly in favour of greater flexibility are, in particular:

- parents, especially mothers, who would prefer to be able to take their children to and from school, and to work around this commitment
- employees, many of them in the older age range, who have elderly or sick relatives who they would like to be more available for.

A more flexible approach would make it easier for many staff to fulfil these kinds of demands on their time.

An initial study questioned nearly 140 employees in a cross-section of ages. A large majority were in favour of a more flexible approach, in particular the women and the younger members of the company. It is worth noting that a minority of staff were against the introduction of flexible working hours. Appendix I gives the full results of this study.

Possibilities

Since this report is looking at the principle and not the detail of a more flexible approach, the options available fall broadly into three categories: retaining the present system, introducing limited flexibility of working hours, and implementing a highly flexible system.

Although the system is not perfect, at least we know it works. The staff all signed their contracts on the understanding that the company worked to standard hours of business, and while it may not be ideal for them it is at least manageable. Better the devil you know.

Implementing any new system is bound to incur problems and expense, consequently retaining the present working hours is the least expensive option in terms of direct cost.

Highly flexible system. A highly flexible system would mean keeping the site open from, say, 7.30 am to 8.00 pm. All staff are contracted to work a certain number of hours a week and time clocks are installed. Employees simply clock on and off whenever they enter or leave the building, until they have reached their full number of hours each week.

This system has the obvious benefit that it can accommodate a huge degree of flexibility which should suit the various demands of all employees. They could even elect to work 35 hours a week spread over only three days. A further benefit to the company would be that doctors' appointments and so on would no longer happen 'on company time' as they do at present. This system does have several disadvantages, however:

- Many staff regard occasional time off for such things as doctors' appointments or serious family crises as a natural 'perk' of the job. With this system they would have to make up the hours elsewhere. Not only would they lose the time off, but many would also feel that the company did not trust them. This would obviously be bad for company morale.
- It would be difficult to implement this system fairly. The sales office, for example, must be staffed at least from 9.00 to 5.30 every day. What if all the sales staff want to take Friday off? How do you decide who can and who can't? What if the computer goes down at 4 o'clock in the afternoon and there are no computer staff in until 7.30 the following morning?

Limited flexibility: This would make asking employees to continue to work an eight hour day, but give them a range of, say, ten hours to fit it into. They could start any time between 8.00 and 10.00 in the morning, so they would finish eight hours later – between 4.00 and 6.00.

On the plus side, this would give the employees the co-operation and recognition of their problems that many of them look for, and would therefore increase staff motivation. For some it would provide a way around their other commitments.

Proposal

Given the number of staff in favour of more flexible working hours, and the importance of staff motivation, it seems sensible to adopt some kind of flexible approach. But it is probably advisable to find a system that allows the significant minority who prefer to stay as they are to do so.

▶

So which is the best system to choose? It is harder to go backwards than forwards in developing new systems: if the highly flexible approach failed it would be difficult to pull back to a less flexible system (in terms of keeping the staff happy). On the other hand, a limited degree of flexibility could easily be extended later if this seemed appropriate.

So at this stage it seems that the most workable system, which contains most of the benefits required by the employees, is the limited flexibility of working hours.

Appendix I

Table of employee responses to the proposal for flexible working hours.

AGE GROUP	MEN Total number consulted	MEN Positive response	MEN Negative response	WOMEN Total number	WOMEN Positive response	WOMEN Negative response
18–30	20	19	1	18	18	0
30–40	23	19	4	29	27	2
40–50	15	8	7	12	8	4
50–60	12	2	10	8	7	1
	70	48	22	67	60	7

TIP Ask yourself whether the report or proposal can be read and understood by someone who knows nothing about the situation, without that person having to ask questions to clarify anything.

Checklist

- Plan reports and proposals very carefully before you start writing.

- Use sub-headings to organise your document into sections.

- Use straightforward, everyday English.

- Avoid jargon, and explain any technical terms that you must use.

- Keep your sentences to 15–20 words, with one main idea in each sentence.

- Use active verbs wherever possible.

- Be clear and concise.

- Write sincerely and personally, as if you are talking to the reader.

- Draft your report first. Some reports will need strict editing, rearranging, rewording, etc.

- Ask someone else to read it through to check that your report meets all these objectives, and that it is readable and accurate.

 TIP Remember to check out The Plain English Guide to Writing Reports in the Appendix. You can find out more about the Plain English Campaign at www.plainenglish.co.uk.

part

Creative and Persuasive Documents

'The most important persuasion tool you have in your entire arsenal is integrity.'

Zig Ziglar

Sometimes in business you need to write a more unusual document, and these commonly fall into creative and persuasive documents. These documents are used in business to advertise or to present information for other parties in a way that helps to persuade or influence the reader in some way.

Effective persuasion means being able to present a message in such a way that will influence others so that they support it. It's essential today that business writers are able to put together a persuasive discussion to influence and convince others. Using a creative style to display the information in a clear, striking and perhaps imaginative format can show your reader that you mean business.

It could be argued that every message must be persuasive. There's an element of truth here, because, just like effective speaking, effective writing must have the act of persuasion at its very core. You must always be 'selling' what you write, even if you're trying to sell yourself.

To be persuasive doesn't demand advanced skills or big words. In fact there are really four things you need to consider before writing your document. Let's look at these four things in a diagram, with the reader (the most important person) in the centre:

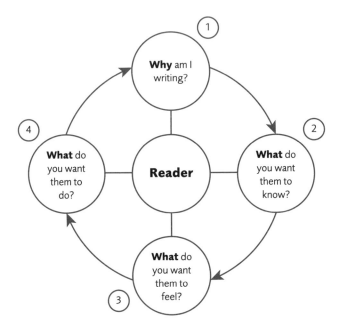

Before writing anything, you must ask yourself:

1 Why am I writing this document? What's my aim?

2 What are the main points I need to tell the reader?

3 What do I want the reader to feel about the topic? You need to touch their emotions.

4 What do I want the reader to do? What are the next steps – action to be taken?

If you answer these questions first, the task of writing your document should be easier and more focused.

21

Complaints and adjustments

Handling complaints

No matter how good our intentions and efforts, there are bound to be occasions when it is necessary to deal with a complaint, or even to make one. Complaints may be necessary for several reasons, such as:

- wrong goods received
- poor service
- unsatisfactory quality of goods
- late delivery
- damaged goods
- prices not as agreed.

Complaining about goods

It's always a good idea to keep your receipt for everything you buy, and then if there's a problem go back to the shop or vendor as soon as possible. Explain what the issue is (being assertive, not aggressive) and say what you expect to be done about it. Hopefully the person in charge will deal with your issue satisfactorily, but if you are still not satisfied you will need to put your complaint in writing. I suggest you should find out the name of the Customer Service Manager so that you can address the letter personally. This is always so much better than sending a general letter or email beginning 'Dear Sirs'.

Complaining about a service

If you are unhappy about service you have received, the supplier deserves a chance to put matters right. If it is necessary to put your complaint in writing, say what you expect to be done and set a deadline. You may wish to consider withholding any further money until the problem has been sorted out satisfactorily, but do check the small print of any contract or credit agreement that you have signed.

Making a written complaint

When you have a genuine complaint you will feel angry, but you must show restraint in your letter, if only because the supplier may not be to blame. Here are some points to remember:

1 Do not delay as this will weaken your position, and the supplier may have difficulty investigating the issue.

2 Do not assume that the supplier is automatically to blame. They may have a perfectly good defence.

3 Don't be offensive or disrespectful. There should be no reason ever to be abusive or impolite, even when complaining about something. Assertiveness without aggression is always the way to go. You don't want to create any ill-feeling that may mean the supplier is unwilling to resolve matters.

4 Here's a good four-point plan for drafting your letter of complaint:

Introduction	If possible, begin with a positive – you might refer to previous good service, for example. Describe the item or service you bought, or the incident that happened. If appropriate, say where and when you bought the item (or when the service was carried out) and how much it cost.
Details	Explain what is wrong, any action you have already taken, to whom you spoke and what happened. This section must be structured logically.
Action	You may form a conclusion here with your feelings about the situation. State what you expect to be done to rectify the situation, for example a refund or repair, or the job done again without charge. Alternatively, you may simply ask the recipient to investigate the matter and take the necessary action.
Close	Close with a simple one-liner saying you hope to receive a prompt reply.

5 Keep copies of your letters or emails, and never send original documents or receipts.

6 Remember to ACE your subject lines when writing a letter of complaint. Your subject line should be:

Appropriate. Use suitable details that are meaningful and relevant.

Concise. Keep it as brief as possible, while giving all the necessary details.

Exact. Make sure all the information is accurate and complete.

 TIP Read more about composing great subject lines in Chapter 6.

Let's look at how this structure works in a real complaint.

A complaint sent by letter

Address the letter personally, and use an appropriate heading ——

This is an ACE subject line

Begin with something positive if possible

State clearly exactly what happened and how you felt

Form a conclusion first and then state what action you expect

Close with a simple one-liner

Dear Mr Stevens

CHUNKY ROASTED VEGETABLE SOUP

I have been shopping at Manley and Simpson for many years, and have always enjoyed the high quality of your food products. However, I was disappointed with my recent purchase of three tins of Chunky Roasted Vegetable Soup, which I bought from your Sheffield High Street branch.

The description of the soup sounded excellent – 'Mediterranean flavours of roasted peppers, courgettes and olives with fusilli pasta'. However, when I opened the first tin I didn't find any pasta in the soup at all. This made the soup quite weak and watery, and not very substantial at all. I am enclosing the label from this tin of soup, and also noted the details on the foot of the can, which said: BB <0827> MRVI 31.10.201-.

I am used to good quality food products from Manley and Simpson, so I was very disappointed that this soup fell far short of my expectations. I felt you would wish to know about this because you will want to address the issue and find out how this happened.

I look forward to your early reply.

Yours sincerely

TIP Confine your complaint to statements of facts and explain your disappointment, rather than showing too much emotion or expressing anger.

Dealing with a complaint

Most suppliers naturally wish to hear if customers have cause to complain. This is better than losing business or the customer taking his/her business elsewhere. It also provides an opportunity to investigate, to explain, and to put things right. In this way goodwill may be preserved. Receiving complaints may also suggest ways in which the supplier's products or services could be improved.

Replying to a written complaint

When dealing with dissatisfied and unhappy customers, remember these guidelines:

- It is often said that the customer is always right. This may not always be the case, but it is sound practice to assume that the customer may be right.

- Acknowledge a complaint promptly. If you are unable to reply fully, explain that it is being investigated and a full reply will be sent later (state a time frame).

- Just as the writer of the complaint must not be offensive or disrespectful, the same must be said of the reply.

- Here's a good four-point plan for drafting your letter of complaint:

Introduction	Personalise your reply and thank the customer. Explain why you appreciate the complaint, and apologise for what has happened.
Details	Acknowledge the customer's point of view. State what action you have taken to fix things. Keep it simple and specific.
Action	Perhaps state what was learned from what happened. Aim to exceed the customer's expectations. It may be appropriate to send a small corporate gift or voucher.
Close	Close positively.

In brief, here's what a good response letter should deliver. It should:

- fix the problem
- fix the relationship
- fix the system or procedures
- ensure something has been learned from what happened.

 TIP Good writers learn to choose their words very carefully and get the tone just right.

Let's look at how the recipient would have replied to the complaint in the letter on page 327:

Reply to complaint

Address the customer by name ——— Dear Miss Stevens

Thank the customer and apologise ——— Thank you for your letter dated 12 May. I am so sorry to hear that your recent purchase of our Chunky Roasted Vegetable Soup did not contain the stated pasta. I do apologise for this, and am glad you brought it to my attention.

Tell the customer how you work hard to ensure quality ——— We try very hard to make sure that all our products are correctly prepared and of the highest quality. It is obvious that on this occasion something went seriously wrong.

Explain what you have done and what will happen now ——— I have passed the details to the department concerned. They are investigating this error immediately and will contact our supplier to make sure this does not happen.

Go the extra mile when appropriate ——— I am pleased to enclose a voucher for £10 as a gesture of our goodwill, and I hope you will continue to be a valued customer of Manley and Simpson.

Yours sincerely

Complaints concerning goods

Complaint concerning wrong goods

If goods are received which are not of the kind or quality ordered then you are entitled to return them at the supplier's expense.

Email complaint

Dear Sirs

Order number and date —— On 12 August I ordered 10 copies of *Background Music* by H Lowery under my order number FT567.

Reasons for dissatisfaction —— On opening the parcel received this morning I found that it contained 10 copies of *History of Music* by the same author. Unfortunately, I cannot keep these books as I have sufficient stock already. I am returning the books by parcel post for immediate replacement, as I have several customers waiting for them.

Action requested —— Please credit my account with the invoiced value of the returned copies and also include my postage cost of £17.90.

Many thanks

Reply

Dear Mr Ramsay

Apologise —— Thanks for your email, and I'm sorry that a mistake was made with your order.

Explain how the mistake happened —— This error happened due to staff shortage during this unusually busy season, and also the fact that these two books by Lowery have identical bindings.

Action taken to rectify the matter —— We have today sent to you 10 copies of the correct title.

I confirm that your account will be credited with the invoiced value of the books and cost of return postage. Our credit note is enclosed.

A closing apology —— Once again, my apologies for this mistake and the inconvenience.

Yours sincerely

Complaint concerning quality

A buyer is entitled to reject goods that are not of the quality or description ordered. However, later deliveries may also not be accepted, even if the goods are correct.

Complaint

Dear Sirs

Reasons for complaint —— We have recently received several complaints from customers about your fountain pens. The pens are clearly not satisfactory, and in some cases we have had to refund the purchase price.

The pens are part of the batch of 500 supplied against our order number 8562 dated 28 March. This order was placed after your representative left a sample pen with us, which we found to be
Further details —— very good quality. We have compared the performance of this sample with that of a number of pens from this batch, and there is little doubt that many of them are faulty – some of them leak and others blot when writing.

The complaints we have received relate only to pens from the batch mentioned. Customers who bought pens before these have always been pleased.

Action required —— We wish to return the unsold balance, amounting to 377 pens, and to have them replaced with pens of better quality.

Close —— Please let us know what arrangements you wish us to make for the return of these unsuitable pens.

Yours faithfully

Reply (accepting complaint)

Dear

Thank you for your email regarding faults in the pens supplied to your order number 8562. We have been very concerned about this, and are glad that you brought this to our notice.

We have tested a number of pens from the production batch you mention, and agree that they are not perfect. The defects have been traced to a fault in one of the machines, which has now been rectified.

Please arrange to return to us your unsold balance of 377 pens. We will be happy to reimburse you for the cost of postage. We have already arranged for 400 pens to be sent to replace this unsold balance. The extra 23 pens are sent with our compliments. You will be able to provide free replacements of any further pens about which you may receive complaints.

Once again, our apologies for this inconvenience.

Best wishes

Alternative reply (rejecting complaint)

If circumstances show that a complaint needs to be rejected, you must show an understanding of the customer's position and carefully explain why a rejection is necessary.

Dear

A tactful opening —— We are sorry to learn from your letter of 10 May that you have had difficulties with the pens supplied to your order number 8562.

Explanation of quality control —— All our pens are manufactured to be identical in design and performance, and we cannot understand why some of them should have given trouble to your customers. It is normal practice for each pen to be individually examined by our Inspection Department before being passed into store. However, from what you say, it would seem that a number of the pens included in the latest batch escaped the usual examination.

Reject the request, but ——— While we certainly understand your concern, we cannot accept
very diplomatically your suggestion to take back all the unsold stock from this batch.
Indeed, there should be no need for this since it is unlikely that
there are a large number of faulty pens. We will gladly replace
any pen found to be unsatisfactory, and on this particular batch
The offer of a discount ——— we are prepared to allow you a special discount of 5% to
adds a 'softener' compensate for your inconvenience.

I hope you will accept this as being a fair and reasonable solution
of this matter.

Please give me a call on 4626123 if you have any further
questions.

Complaint concerning quantity

Surplus goods delivered

When a supplier delivers more than the quantity ordered, the buyer is legally entitled to reject either all the goods or only the excess quantity. Alternatively all the goods may be accepted and the excess paid for at the same rate. In this email, the buyer rejects the surplus goods but is not obliged to return them; it is the supplier's responsibility to arrange for them to be collected.

Dear Sirs

Thank you for your promptness in delivering the coffee we ordered on 30 July. However, 210 bags were delivered this morning instead of 120 as stated on our order.

Our present needs are completely covered and we cannot make use of the 90 bags sent in excess of our order. We will therefore hold these extra bags in our warehouse until we receive your instructions.

We hope to hear from you soon.

Shortage in delivery

When a supplier delivers less than the quantity ordered the customer cannot be forced to accept delivery by instalments. The customer could insist on immediate delivery of the balance.

Dear Sirs

Thank you for so promptly delivering the gas coke ordered on 20 March. Although we ordered 5 tonnes in 50 kg bags, only 80 bags were delivered. Your carrier was unable to explain the shortage and we have not received any explanation from you.

We still need the full quantity that we ordered, so please arrange to deliver the remaining 20 bags urgently.

I look forward to your confirmation.

Complaint to manufacturer

Customer's complaint

In this letter the buyer was advised by the supplier to write directly to the manufacturer regarding faulty goods.

Dear Sirs

On 21 September I bought one of your 'Big Ben' alarm clocks from Stansfield's Jewellers in Leeds. Unfortunately, I have been unable to get the alarm system to work and am very disappointed with my purchase.

The manager of Stansfield's has asked me to return the clock to you so that the fault can be corrected. This is enclosed.

Please arrange for the clock to be put in full working order and return it to me as soon as possible.

If you have any queries, please call me on 62387282.

Yours faithfully

Manufacturer's reply

In this reply the manufacturer shows genuine concern about the complaint and does everything possible to ensure customer satisfaction. The considerate manner in which the complaint is treated helps to build a reputation for reliability and fairness.

Dear Mrs Wood

Thank you for your letter of 20 September enclosing the defective 'Big Ben' alarm clock.

I'm sorry you have had such problems with this clock, and I have passed it to our engineers for inspection.

We are arranging to replace your clock with a new one that has been tested thoroughly to ensure that it is in perfect working order. This will be sent to you within the next few days.

I am confident that the replacement clock will prove satisfactory and give you the service you are entitled to expect from our products.

Once again, our apologies for the trouble and inconvenience this has caused you,

Yours sincerely

CAUTION It's easy to cause offence if you use the wrong tone. When writing and responding to complaints, you must take great care with your tone. Read more about tone in Chapters 3 and 16.

Complaints about delivery

No supplier likes to be accused of negligence or carelessness, which is often what a complaint about packaging amounts to. Such complaints must be carefully worded so as not to give offence. Nothing can be gained by using sarcasm or insults. Courtesy is more likely to get you the results you want. When writing such letters or emails, show that you don't like having to complain, but explain that the trouble is too serious not to be reported.

Complaint concerning damaged goods

Complaint

The writer of this email points out damage that was discovered after checking the consignment. The writer tactfully avoids making any suggestion that the damage is due to faulty packing.

Subject: Our order number R569

Dear Sirs

Introduction and background details — The 210 compact discs that we ordered on 3 January were delivered yesterday. Unfortunately, 18 of them were badly scratched.

Explain details which evolved after receipt of goods — The package containing these goods appeared to be in perfect condition and I accepted and signed for it without question. It was when I unpacked the contents that I noticed the damaged CDs. I can only assume that this was due to careless handling at some stage prior to packing.

Attach full list of damaged goods and request replacement — I am attaching a list of the damaged goods and hope you will replace them soon. They have been kept aside in case you need them to support a claim on your suppliers for compensation.

I look forward to your prompt reply.

Reply

The supplier's reply complies with the customer's request and shows a desire to improve the service to customers.

Dear James

Acknowledge letter and show regret about damages — I am sorry to learn from your email that some of the compact discs supplied were damaged.

Give details about replacements — We have sent replacements for the damaged goods by parcel post this morning. It won't be necessary for you to return the damaged goods. They may be destroyed.

Give further information about follow-up action — Despite the care we take in packing goods, there have recently been several reports of damage. To avoid further inconvenience to customers, as well as expense to ourselves, we are now seeking the advice of a packaging consultant in the hope of improving our methods of handling.

Assurance about future orders — We apologise once again for this, and hope the steps we are taking will ensure the safe arrival of all your orders in future.

Best wishes

Complaint about bad packing

Complaint

Subject: Our order number C395

Dear Sirs

Introduction about reason for writing — The carpet supplied to our order dated 3 July was delivered by your carriers this morning.

Details about complaint — We noticed that one of the outer edges of the wrapping had been worn through, presumably as a result of friction in transit. When we took off the wrapping, we found that the carpet itself was soiled and slightly frayed at the edge.

Further details and questions about precautions — This is the second time in three weeks that we have had to write to you about the same matter. We find it hard to understand why precautions could not be taken to prevent a repetition of the earlier damage.

Suggestions about future handling of orders — Although other carpets have been delivered in good condition, this second experience within such a short time suggests the need for special precautions against friction when carpets are packed onto your delivery vehicles. We hope you will bear this in mind in handling our future orders.

Requests for special concession — In view of the condition of the present carpet we cannot offer it for sale at the normal price and propose to reduce our selling price by 10%. We suggest that you make us a similar allowance of 10% on the invoice cost. If you cannot do this, we shall have to return the carpet for replacement.

I hope to hear from you soon.

Reply

Dear

Apologise for customer's dissatisfaction ——— I am sorry to hear that this carpet was damaged on delivery.

Explain circumstances surrounding the issue ——— Our head packer informs us that the carpet was first wrapped in heavy oiled waterproof paper and then in a double thickness of jute canvas. Under normal conditions this should have been enough protection. However, on this occasion our delivery van contained a full load of carpets for delivery to other customers on the same day, and it is obvious that special packing precautions are necessary in such cases.

Mention follow-up action taken ——— In all future consignments, we are arranging for specially reinforced end-packings. This should prevent any future damage.

Confirm special discount to customer ——— We realise the need to reduce your selling price for the damaged carpet and agree to the special allowance of 10% which you suggest.

Once again, I'm sorry for the inconvenience.

Complaint regarding non-delivery

Complaint

Subject: Our Order RT56

Dear

On 25 September we placed an order for headed notepaper and invoice forms. You acknowledged the order on 30 September. We have still not received advice of delivery, so we are wondering if our order has been overlooked.

Your representative promised an early delivery and this was an important factor in persuading us to place this order with you.

This delay is causing considerable inconvenience, so we hope you can complete the order immediately, otherwise we shall have no option but to cancel it and obtain the stationery elsewhere.

I hope to hear from you soon.

Reply

Only a very tactful can keep the goodwill of this customer, who is obviously feeling very let down. With an understanding and helpful reply from the printer as shown here, the customer will surely not continue to feel annoyed.

> Dear Mr Sargeant
>
> Thank you for your email, and I quite understand your annoyance at not having received the stationery ordered on 25 September.
>
> Orders for printed stationery are taking from 3 to 4 weeks for delivery, and our representatives have been instructed to make this clear to customers. Apparently you were not told that it would take so long, and I am sorry for this oversight.
>
> I have immediately checked into this, and am pleased to tell you that your stationery will be despatched tomorrow by express parcel post. It should reach you within 24 hours.
>
> Once again I'm sorry there has been this misunderstanding, and hope you are pleased with the goods when they are received.
>
> Best wishes

Complaint regarding frequent late deliveries

This correspondence shows how important it is when sending letters of complaint to write with restraint and not to assume that the supplier is at fault.

Complaint

> Dear Mr Johnson
>
> We ordered 6 filing cabinets from you on 2 July on the understanding that they would be delivered within one week. However, these have only been received this morning, which is 5 days later than expected.
>
> We hope you can appreciate the problem this causes, because we cannot give delivery dates to our customers unless we can rely on undertakings given by our suppliers.
>
> Unfortunately, there have been delays on several previous occasions, and because of their increasing frequency we don't feel that business between us can continue in such conditions.
>
> We hope you will understand our position in this matter, and would like your assurance that we can rely on punctual delivery of our orders in future.
>
> I hope to hear from you soon.

Reply

In this reply the supplier carefully explains that the fault is not on their part, and goodwill with the customer should be retained.

Dear

Thanks for your message regarding delays in delivery. This actually came as a surprise as we have not received any earlier complaints from you. This naturally led us to believe that goods supplied to your orders were reaching you promptly.

It is our usual practice to deliver goods well in advance of the promised delivery dates. The filing cabinets that you ordered on 2 July left us on 8 July, so we are very concerned that our efforts to ensure punctual delivery have been frustrated by delays in transit. It is possible that other customers are also affected, so we are taking up this whole issue with our carriers.

Thank you for drawing this situation to our attention, and please accept our apologies for the inconvenience.

Best wishes

Complaint regarding uncompleted work

This correspondence relates to a builder's failure to complete work on a new bungalow within the agreed contract time. The buyer's letter is firm but reasonably worded. The builder's reply shows understanding and is convincing, businesslike and helpful.

Complaint

Dear Mr Jones

BUNGALOW AT 1 CRESCENT ROAD, CHINGFORD

When I signed the contract for the building of this property you estimated that the work would be completed and the bungalow ready for occupation in about six months' time. That was two months ago and the work is still only half finished.

The delay is causing inconvenience not only to me but also to the person buying my present home, which I cannot transfer until this bungalow is finished.

I hope you can proceed with this work without any further delay. Please let me know when you expect it to be completed.

Yours sincerely

Reply

Dear Mr Watson

BUNGALOW AT 1 CRESCENT ROAD, CHINGFORD

Thank you for your letter of 18 June. We are of course aware that the estimated period for completion of your bungalow has already been exceeded. We realise that this delay must be causing you considerable inconvenience.

Unfortunately, you will recall that we had an exceptionally severe winter. Work on your property was quite impossible during several prolonged periods of heavy snow. There has also been a nationwide shortage of building materials, especially bricks and timber, from which the trade is only just recovering. Both of these difficulties could not have been foreseen. Without them, the estimated completion period of six months would have definitely been met.

Now that the weather has improved, work is now going ahead very well. Unless we have other unforeseen delays, we can safely promise that the bungalow will be ready for you by the end of August.

Yours sincerely

 TIP If you wouldn't say it, don't write it. Today's writing should be conversational, as if you are speaking to the reader.

Complaint regarding delivery charges

Some customers are only too ready to complain if things do not suit them. Others who are dissatisfied do not complain, but instead they quietly withdraw their custom and transfer it to some other supplier. This correspondence relates to such a case.

Supplier's enquiry

Dear Sirs

We are sorry to notice that we have had no orders from you since last April. As you have not informed us of defects in our products or the quality of our service, we can only assume that we have given you no cause to be dissatisfied. If we have, then we hope you will let us know.

If the cause of your discontinued orders is the present depressed state of the market, you may be interested in our attached price list showing a reduction of 7.5% on all grocery items.

If we have given you any reason to be dissatisfied, we hope you will give us the opportunity to put it right so that our custom can be renewed.

I hope to hear from you soon.

Customer's reply (complaint)

Dear

Thank you for your email. As you wish to know why we have not placed any orders with you recently, I will point out a matter that caused us some annoyance.

On 21 April last year we sent you two orders, one for £274 and one for £142. Your terms at the time provided for free delivery of all orders for £300 or more, but although you delivered these two orders together we were charged with the cost of carriage.

As the orders were submitted on different forms, we realise you had a perfect right to treat them as separate orders. However, for all practical purposes they could very well have been treated as one, as they were placed on the same day and delivered at the same time. The fact that you did not do this seemed to be a particularly unfair way of treating a regular long-standing customer.

I look forward to receiving comments on this matter.

Supplier's reply

Suitable introduction ——

Circumstances ——
surrounding the
situation are explained
in detail

Further details given to ——
assure the customer that
this situation will not be
repeated

This tactful close ——
expresses a hope for
renewed business
dealings

> Dear
>
> Thank you for your message. I'm so glad you have given us the opportunity to explain a most unfortunate misunderstanding.
>
> Our charge for carriage on your last two orders arose because they were for goods dealt with by two separate departments, neither of which was aware that a separate order was being handled by another.
>
> At that time these departments were each responsible for their own packing and despatch arrangements. This work has recently been taken over by a centralised packing and despatch department, so a repetition of the same kind of misunderstanding is now unlikely.
>
> In the circumstances, I hope you will feel able to renew your former custom, and we look forward to working with you again.
>
> Best wishes

Cancelling orders

A buyer is legally entitled to cancel his/her order at any time before it has been accepted by the supplier, or if:

- the goods delivered are of the wrong type or quality (if they do not conform to sample)
- the goods are not delivered by the stated time (or within a reasonable time if no delivery date has been fixed)
- more or less than the quantity ordered is delivered
- the goods arrive damaged (but only where transportation is the supplier's responsibility).

Unless the contract provides otherwise, it is the buyer's legal duty to collect and transport the goods from the supplier's premises. This would be so where the goods are sold ex works or similar terms. The buyer is then liable for any loss or damage that happens during transport. Similarly under an *fob* or a *cif* contract, the customer is liable from the time the goods are loaded onto the ship.

Buyer asks to cancel order due to adequate stocks

Customer's email

Dear Sirs

On 2 March I ordered 100 tennis rackets to be delivered at the end of this month.

As persistent bad weather has seriously affected my sales, I feel that my present stock will probably satisfy this season's demand. Therefore, I would like to cancel part of my order and ask you to deliver only 50 of these rackets instead of the 100 ordered.

I am sorry to make this request so late but hope that you will be able to agree to it in view of our long-standing business relationship. If sales improve I will get in touch with you again and take a further delivery.

I hope to hear from you soon.

Supplier agrees to cancel order

A supplier will often agree to cancel or modify the buyer's order for a number of reasons:

- a wish to oblige a good customer
- the loss of profit involved may be minimal
- it helps to create customer goodwill
- there may be a ready market for the goods elsewhere
- the customer's financial position may be doubtful
- legal proceedings are costly.

Dear

Thanks for your email. We are naturally disappointed that you should need to cancel half of your order. However, we always like to oblige our regular customers and in the circumstances we are prepared to reduce the number of rackets from 100 to 50 as requested.

We do hope your sales will improve enough so that you can take up the balance of your order at a later date.

All the best

Supplier refuses to cancel order

The supplier will sometimes refuse to cancel an order for various reasons:

- a wish to retain a sale
- the manufacture of goods that cannot easily be sold elsewhere may have begun
- a keen entrepreneur may be unwilling to forgo their legal rights.

The message refusing a request for cancellation must be worded carefully and considerately if it is not to cause offence and drive a customer away. It's important to show that you understand the buyer's challenges, and tactfully explain the difficulties that cancellation would create. The reasons given must be convincing, otherwise the supplier is liable to lose the customer's goodwill.

Dear

Thanks for your message asking us to cancel part of your order for tennis rackets.

We are sorry you find it necessary to make this request, especially at this late stage. To be able to meet our customers' needs promptly we have to place our orders with manufacturers well in advance of the season. When estimating quantities, we rely very largely upon the orders we have received.

We do not like to refuse requests of any kind from regular customers. However, we feel we have no choice but to do so. All orders, including your own, have already been made up and are awaiting delivery.

I hope you will understand why we must hold you to your order. If we had received your request earlier we should have been glad to help you.

Best wishes

Cancellation of order through delay in delivery

Subject: Order number 8546 dated 18 August

Dear

When we placed this order we stressed that we needed delivery by 4 October at the latest.

We have already written to you twice reminding you of the importance of prompt delivery. However, as you have been unable to deliver these goods on time, we are left with no choice but to cancel the order.

These goods were required for shipment abroad, and as the boat by which they were to be sent sails tomorrow, we have no means of getting them to our client in time for the exhibition for which they were required.

Please acknowledge this cancellation.

Personal complaints

There may be many reasons why it is necessary for you to write a letter or email of complaint. Although you will probably feel angry, you must remember that the other party might have a good explanation or may not be to blame. Confine your letter to a statement of the facts, stress your disappointment, then ask what the organisation intends to do.

Missing mileage request to airline

Customer's email

Subject: Missing mileage request

Dear Sirs

I attach my completed form for missing mileage from my Axon Car Rental in July this year. Also attached is a copy of the missing mileage request form faxed to you last week by my travel agent (Travel Shop) who booked this flight for me.

I hope you are able to credit me with the missing miles. This flight back to the UK was arranged very quickly because I received a phone call to say that my mother was ill and was being taken into hospital. I arranged the flight and flew home very quickly. I remember my travel agent telling me that if I paid a few hundred dollars extra, my flight would be eligible for air miles, so I chose to do this. However, when I got to the check-in desk at the airport, I was very upset and anxious, and I don't recall reminding the check-in clerk that I was a Content Club member. Perhaps this is why my account has not been credited with the extra miles.

As you will see from my Axon Car Rental receipt, I did mention that I was a Content Club member when I picked up my hire car at Manchester airport. However, again these miles have not been credited to me.

I hope you will be able to credit my account with the air miles that are due from both my air travel and my hire car rental.

I look forward to your early reply.

Yours faithfully

Reply from the airline

Dear Miss Turner

Thank the customer for ——— Thank you for your email, and we are sorry for the printing error
the letter and apologise on your recent Content Club statement.
up front

Our Information Management team has isolated the error and the
Explain what has been ——— problem happened at the printing stage. However, I can assure
done about the situation you that your personal data has not been affected and the
information stored on the Content Club database is correct.

I am attaching your revised Content Club statement. If you have
Give additional ——— any further questions about this, please contact me. Alternatively
information to reassure your most recent travel history is available by logging onto our
the customer website **www.contentclub.com.sg**.

Go the extra mile ——— To make up for this inconvenience, we have also credit your
account with 1,000 bonus air miles.

Thank you for your understanding.

 TIP Note how the writer says thank you in the first paragraph.
Many people think you should not say thank you to someone who
has complained. But why not? They have given you a chance to put
something right and retain goodwill.

Complaint about insurance claim

Customer's letter

Dear Mr Watson

CLAIM AL54323432 – STORM DAMAGE TO ROOF

I received a cheque for £623 dated 26 January in payment of my recent claim.
However I wish to place on record how much upset has been caused by the way your
Claims Assessor, Mr Michael Tan, handled this claim.

When Mr Tan first called me he specifically told me that he believed I had been overcharged for this work, and said he would expect to pay that price for work on a double garage rather than a single garage like mine. Mr Tan said that in his opinion I should neither use nor recommend this contractor again. He then told me it was unlikely that I would receive payment for the full cost that I had paid out. He never mentioned that the reason for not receiving full payment was because of the nature of my insurance policy.

Consequently, I wrote to the contractor, Mr Lance Ashe, to complain about his pricing, stating that I was very upset thinking that he could have knowingly taken advantage by overcharging a 73-year-old woman. Mr Ashe called me immediately and explained his charges in detail, as he was very upset that he had caused me some distress. I believe Mr Ashe then called Mr Tan, because he later reported back to me that Mr Tan had told him that the reason I would not receive full payment in regard to my claim was because of the type of policy that I hold, which does not cover wear and tear. This was the first time this issue had been brought to my attention, so you can imagine my surprise.

This situation was only explained in Mr Tan's letter of 2 February. If this had been explained to me in the first place, I would have been able to accept it and would certainly not have questioned Mr Ashe's charges. Instead, by telling me initially that I had been overcharged for this work, it caused a great deal of upset not only for me but also for Mr Ashe who was naturally most upset that anyone should think his work was unfairly priced.

I believe this claim was handled badly by Mr Tan from the beginning when he led me to believe that I would not be reimbursed in full because I had been overcharged – not because of the nature of my policy, which I now know to be the case. This has caused me a great deal of embarrassment and upset, not to mention the upset between me and Mr Ashe.

I hope you will look into this and ensure that such claims are handled more appropriately in the future.

Yours sincerely

Acknowledgement from insurance company

The insurance company cannot reply immediately as they need to investigate. However, it is good business practice to send a brief note explaining what is happening.

Dear Mrs Richardson

Thank you for your letter of 4 February. I am sorry to learn of the problems you have experienced recently with your claim.

I am looking into this issue immediately, and I will be in touch again within the next week. Meanwhile, if you would like to discuss this matter with me, please give me a call on 0114 2347827.

Yours sincerely

Detailed reply from insurance company

Dear Mrs Richardson

Following my letter of 6 February, I have reviewed our file on your recent claim.

I am sorry that you feel you were given the impression by our claims assessor that there was a problem with the work carried out by your chosen contractor and the price he charged. It was not the assessor's intention to distress you with his initial thoughts, and I am sorry about the concern this caused you. However, after discussion with Mr Ashe, the assessor was satisfied that the cost was reasonable and that the work had been completed satisfactorily.

Our assessors are highly trained to investigate and ensure that claims settlements are fair and reasonable. They have a responsibility to ensure that all relevant enquiries are made before settling a claim, and I am happy that the assessor has acted properly in this respect. However, it is unfortunate that in the first place he gave you the wrong impression, and I do apologise for this.

We are always interested in any feedback from our customers on the service that we provide, and are continually looking for ways to improve this.

Thank you again for writing to us. If I can be of any further help, please let me know.

Yours sincerely

The wrong way and right way for writing a complaint

The wrong way

This letter has obviously been written in anger and in haste. The tone is very disrespectful, the contents are not structured logically, important details have been omitted, and the letter would not achieve anything.

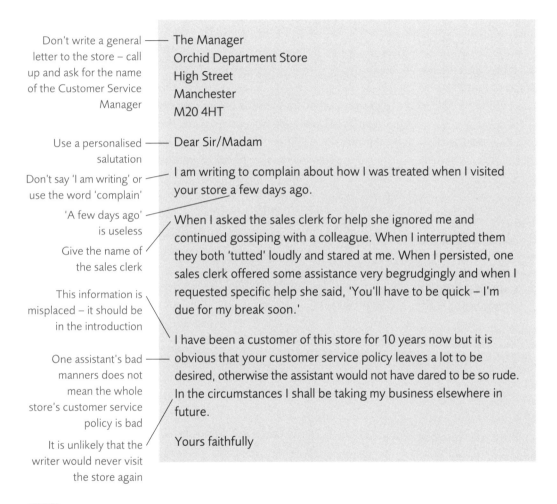

Don't write a general letter to the store – call up and ask for the name of the Customer Service Manager

The Manager
Orchid Department Store
High Street
Manchester
M20 4HT

Use a personalised salutation

Dear Sir/Madam

Don't say 'I am writing' or use the word 'complain'

I am writing to complain about how I was treated when I visited your store a few days ago.

'A few days ago' is useless

Give the name of the sales clerk

This information is misplaced – it should be in the introduction

When I asked the sales clerk for help she ignored me and continued gossiping with a colleague. When I interrupted them they both 'tutted' loudly and stared at me. When I persisted, one sales clerk offered some assistance very begrudgingly and when I requested specific help she said, 'You'll have to be quick – I'm due for my break soon.'

One assistant's bad manners does not mean the whole store's customer service policy is bad

It is unlikely that the writer would never visit the store again

I have been a customer of this store for 10 years now but it is obvious that your customer service policy leaves a lot to be desired, otherwise the assistant would not have dared to be so rude. In the circumstances I shall be taking my business elsewhere in future.

Yours faithfully

 CAUTION When replying to a letter of complaint, do not be rude or sarcastic. This will cause ill-feeling, which will be counterproductive.

The right way

This letter is more structured, the tone is respectful, and the writer remains courteous while giving a clear indication of the reason for her dissatisfaction. All the details are provided so that the manager will have no problem in investigating the matter.

Use a personalised salutation ———— Dear Mrs Williams

It's always good to begin with a compliment when writing a complaint ———— I have been a customer of Orchid Department Store for the last 10 years and have always been very happy with the service.

Give all the details – where, when, who, what time, what happened ———— However, when I visited the Ladies Department around noon on Monday 12 June, one of your assistants, Sandra Wong, was very unprofessional. When I asked for help she continued talking to her colleague. Eventually she said abruptly, 'You'll have to be quick – I'm due for my break soon!'

Don't tell the manager what you expect – just ask her to investigate ———— I was very surprised at this, as it is not the sort of service that I have come to expect from staff of Orchid, and I am sure you will investigate this matter.

Finish with a simple close stating that you expect an early reply ———— I hope to hear from you soon.

Yours sincerely

An unsatisfactory reply to a complaint

Here is a reply to a complaint that has been written quickly and in a tone that is far from courteous.

Subject: Your Complaint

Your complaint about your fax machine that you bought from us last year has been passed to me for my attn. Please be informed that your policy document shows that you only have a one year guarantee for this product and it ran out on 2nd Sept. So if you want it fixing you will have to pay for it.

Let me know what you want to do.

Here is the same reply written in a more courteous tone.

Subject: ST101 Fax Machine

Dear Robin

Thank you for your message. I am sorry to hear about the problems you have experienced with this fax machine.

I have checked your policy document and unfortunately our one-year guarantee for this machine ended on 2 September. I am sorry to say that you must bear the cost of any repairs.

We will be happy to repair the machine for you and can promise immediate attention and reasonable terms.

Please give me a call soon on 2874722 to discuss this.

Grace Peng
Global Communications
www.global.co.cn

Useful expressions

Letters of complaint

Openings

- On ___ we placed an order for ___.
- The goods we ordered from you on ___ have not yet been delivered.
- We have not yet received the goods ordered on ___.
- One of the cases of your consignment was badly damaged when delivered on ___.
- We have received several complaints from customers regarding the goods supplied by you on ___.

Useful central phrases

- I am very unhappy with ___.
- This situation is causing us a lot of inconvenience.
- This standard of workmanship is not what I expect from you.
- This service is well below the standard expected.
- I felt you would wish to know about this.
- I know you will wish to look into this and find out what happened.

Action/close

- Please look into this matter and let us know the reason for this delay.
- We hope to hear from you soon that the goods will be sent immediately.
- I hope you will investigate this and take the necessary action.
- We feel there must be some explanation for this delay.
- In the circumstances, I hope to hear that you are prepared to make some allowance.
- I hope to receive a complete refund soon.

Replies to complaints

Openings

- I am sorry to hear that the goods sent under this order did not reach you until ___.
- I am sorry that you have experienced delays in the delivery of ___.
- Thank you for your letter of ___ and we are sorry for this most unfortunate mistake.
- Thank you for your message and we are sorry for the unfortunate mistake that was made.

Useful central phrases

- We appreciate the opportunity to clarify this issue.
- It is obvious that on this occasion a mistake happened on our side.
- You have rightly pointed out that . . .
- In the circumstances, it is important that we make amends for your inconvenience.
- Due to an oversight . . .
- It is unfortunate that . . .
- I am sorry about the distress this caused you.

Action/close

- We assure you that we are doing all we can to rush delivery. Once again, we apologise for the inconvenience this delay is causing you.
- We hope you will be satisfied with the arrangements we have made.
- We hope these arrangements are satisfactory and look forward to receiving your future orders.
- Once again, we apologise for the inconvenience.
- We are sorry for the unfortunate mistake and can assure you that a similar incident will not happen again.
- As a gesture of goodwill, I am pleased to enclose ___.
- Thank you once again for taking the time to write to us.

Checklist for making a complaint

■ Act promptly.

■ Show restraint in your wording – the supplier may have a good defence.

■ State the facts briefly, exactly and clearly.

■ Avoid rudeness.

■ Suggest desired results/action.

Checklist for responding to a complaint

■ Investigate the matter promptly.

■ Show understanding and empathy.

■ If unreasonable, be firm but polite, and try not to offend.

■ If you are at fault, admit it and apologise

■ Explain how the matter will be put right or has been rectified.

■ Never point fingers or blame staff.

■ Provide extra effort, information or compensation if appropriate.

■ Give personalised attention.

■ Thank the person again for writing.

■ Reassure the customer of future good service, to build loyalty.

Goodwill messages

One of the most important functions of all communications is to create good business relations. Many managers and executives take the opportunity to send goodwill messages on various occasions such as:

apologies	unwelcome news	sympathy	welcome
promotion	congratulations	death	special award
thanks	condolence	appreciation	wedding

Every opportunity should be taken to write goodwill messages. They are appreciated by customers and colleagues and are very good for business. For little cost and effort, they not only strengthen existing relationships but may also create new business opportunities.

It's essential that goodwill messages are written and sent promptly. They should be brief and to the point, always sincere and informal. Handwritten notes will give an added touch of sincerity and intimacy where appropriate.

General goodwill messages

The following messages are examples of ways in which goodwill can be built into the everyday business letter or email. The tone is courteous and friendly, and the added touches of personal interest are certain to make a great impression.

 TIP Your writing reflects an impression of you – make sure it's a good one.

Email with short personal greeting

A personal touch may sometimes take the form of a short final paragraph conveying a personal greeting.

Dear Mr Ellis

I am sorry not to have replied sooner to your letter of 25 October regarding the book *English and Commercial Correspondence*. My Export Director is in Lebanon and Syria on business. As I am dealing with his work as well as my own, I am afraid my correspondence has fallen behind.

Whether this book should be published in hardback or paperback is a decision I must leave to my Editorial Director, Tracie James. I have forwarded your message to Tracie, and I'm sure she will contact you very soon.

I hope you are keeping well.

With best wishes

Email with extended personal greeting

An even more personal note may be introduced in the final paragraph.

Dear Marion

Thanks for sending me your new manuscript recently. I've now had an opportunity to review this.

Your book presents a concise and clear account of the new import regulations, with good examples of how they are likely to be applied. I have made some more detailed comments on my written review, which is attached.

I remember you mentioned that you will be spending your summer holiday in the south of France. I'm sure it will be a great break for all the family. I hope you have good weather and an enjoyable time.

Safe trip!

John Ellis

Email explaining delayed reply

It's always best to reply to an email on the day it is received. If this is not possible, it's good to send an acknowledgement on the same day, with a note about when the sender can expect a proper reply from you.

Dear Mrs Jones

Thanks for your enquiry today. I am sorry we cannot immediately send you the catalogue and price list you requested, as we are presently waiting for new stocks.

I'm expecting the printers to deliver new stocks early next week, and I will send a copy to you immediately I receive them.

Sorry for this delay.

Jim Fong

Supplier's email with friendly tone

Customers always look for a spirit of friendliness in those with whom they pursue a business collaboration. In this message the writer is both helpful and friendly. The aim is to interest the prospective customer, to create a feeling of confidence, and to win their consideration, friendship – and ultimately their custom.

Dear John

Many thanks for your email, and I am pleased to enclose our catalogue and price list.

I'm sure you will notice a lot of changes in our new catalogue, which we feel is both more attractive and informative. We've also shown particulars of our trade discounts inside the front cover, so it's very easy for reference.

Next time you are in Bristol, it would be great to meet you. I would be happy to show you our factory where you could see for yourself the high quality of materials and workmanship put into our products. You can then see all our latest fancy leather goods, and you'll return home with interesting and useful information for your customers.

Please give me a call if I can help you with anything. My number is 07811 2874828.

All the best

Mary Jones

TIP Remember, the aim of all your communications today must be to build relationships.

Message welcoming a visitor from abroad

When customers from overseas visit your country, it is sound business practice to extend hospitality and to give any help and advice you can. The tone of such messages must be sincere and friendly, giving the impression that the writer is genuinely keen to be of service.

Dear Richard

Thanks for your email today, and I'm delighted to hear that your colleague, Mr John Gelling, is making plans to visit England in July. We shall be very pleased to welcome him and to do all we can to make his visit enjoyable and successful.

I understand this will be John's first visit to England, and I am sure he will be travelling around making contact with lots of potential clients. I do have some suggestions, and would be happy to introduce him to several firms with which he may like to do business.

Perhaps you can pass on my email address to him, so that we can be in touch personally. I'd also be happy to arrange to meet him at the airport and drive him to his hotel. He can be assured of a warm welcome.

Best wishes

Ron Marshall

Letters of apology

When it is necessary to apologise for something, it's important to get the tone right. Sometimes you may have to swallow your pride and say you are sorry even if you're not. Legal pressure may mean an apology is necessary if you have caused injury or offence to someone.

Apology for poor service

Dear Mrs Taylor

Background details regarding complaint — Thank you for your letter of 12 June regarding the poor service you received when you visited our store recently.

State action taken and express regret — The incident was most unlike our usual high standards of service and courtesy. The member of staff who was rude to you has been reprimanded. He also feels very bad about his outburst, and sends his personal apologies.

Follow-up action — I am enclosing a gift voucher for £20 which you may use at any Omega store.

If I can be of any further help to you, please let me know.

Apologise again — With my apologies once again.

Stephen Chong
Customer Service Manager

Apology for cancelling an appointment

Dear Michael

I am so sorry I had to cancel our meeting yesterday at such short notice. As my secretary explained to you, unfortunately an urgent matter came up that I had to deal with immediately.

I believe our appointment has been rearranged for next Tuesday 12 May at 11.00 am. I shall look forward to seeing you then, and hope we can extend our meeting over lunch.

Best wishes

Martina Fletcher

 TIP Remember, how you say something is often more important than what you say. Consider tone carefully.

Messages in which tone is particularly important

In business it is sometimes necessary to refuse requests, increase prices, explain an unfortunate oversight, apologise for mistakes, etc. In such letters, tone has to be the writer's main concern. Without due consideration, you could easily cause offence, bad feeling could be created, and business may be lost.

Letter conveying unwelcome news

It is sometimes necessary to refuse a request or to convey unwelcome news. Remember that it's always good to think of the reader when you write. Prepare the way for their disappointment by starting with a suitable opening paragraph, and use an appropriate tone.

Dear Mr Foster

It was good of you to let me see your manuscript on *English for Business Studies*. I read it with interest and was impressed by the careful and thorough way in which you have treated the subject. I particularly like your clear and concise style of writing.

Had we not recently published *Practical English* by Freda Leonard, a book that covers very similar material, I would have been happy to accept your manuscript for publication. However, in the circumstances, I am unable to do so. I'm returning your manuscript with this letter.

I am sorry to have to disappoint you, but I feel sure that you will have much success with another publisher. Good luck!

Yours sincerely

Letter disclaiming liability for loss

Here is another letter in which the opening paragraph is used to prepare the recipient for the rejection of his insurance claim.

Dear Mr Burn

I was sorry to hear in your letter dated 23 November about the recent fire in your warehouse. As soon as we heard from you, we sent a representative to inspect and report on the damage.

We have now received the report and it confirms your claim that the damage is extensive. However, it also states that a large proportion of the stock damaged or destroyed was very old and some of it obsolete.

Therefore, we cannot accept your figure of £45,000 as a fair estimate of the loss, as this appears to be based on the original cost of the goods.

Please contact me soon to discuss a more appropriate figure for this claim.

Yours sincerely

Letter refusing a request for credit

A letter refusing a request for credit without causing offence is one of the most difficult to write. Refusal will be prompted by doubts about the would-be creditor's standing, but the letter must contain no suggestion of this. It's important to be very tactful and to give other reasons for the refusal.

This letter is a wholesaler's reply to a trader who has started a new business that appears to be doing well. However, the business has not been established long enough to inspire confidence in the owner's financial standing.

Dear Miss Wilson

We were pleased to learn of the good start of your new business, and glad that you approached us with a view to placing an order.

The question of granting credit for newly established businesses is never an easy one. Many owners get into difficulties because they overcommit themselves before they are thoroughly established. Although we believe that your own business has great promise, we feel it would be better for you to make your purchases on a cash basis at present. If this is not possible for the full amount, we suggest that you cut the size of your order, perhaps by half.

If you are willing to do this, we will allow you a special cash discount of 4% in addition to our usual trade terms. If you agree with this suggestion, the goods could be delivered to you within three days.

We hope you will look upon this letter as a mark of our genuine wish to enter into business with you on terms that will bring lasting satisfaction to us both. When your business is firmly established we will be very happy to welcome you as one of our credit customers.

Yours sincerely

CAUTION You never get a second chance to give a good first impression. Get it right before you click 'send'.

Email apologising for an oversight

If you have made a mistake or are in any way at fault, it's always best to admit it freely and without excuses. A message written in an apologetic tone is likely to create goodwill, and it will be difficult for the recipient to continue to feel a grudge against you.

Dear Mrs Wright

I was very concerned when I received your email today stating that the central heating system in your home has not been completed by the date promised.

On referring to our earlier correspondence it seems that I had mistaken the date for completion. The fault is entirely mine and I am very sorry about this.

I realise how much inconvenience my oversight must be causing you and will do everything possible to avoid any further delay. I have already given instructions for this work to take first priority. Our engineers will be placed on overtime to complete the work. These arrangements should ensure that the work is completed by next weekend.

My apologies once again for the inconvenience. Please call me personally if you have any other questions.

Best wishes

Charles Long

Letter regretting price increase

Customers will naturally resent increases in prices of goods, especially if they feel the increases are not justified. You can preserve goodwill by explaining clearly and convincingly the reasons for the increases.

Dear

Many businesses have been experiencing steadily rising prices over the past few years and it will come as no surprise to learn that our own costs have continued to rise with this general trend.

Increasing world demand has been an important factor in raising the prices of our imported raw materials. A recent national wage award has added to our labour costs, which have been increased still further by constantly increasing overheads.

Until now we have been able to absorb rising costs by economies in other areas. Unfortunately, we are no longer do so, and therefore have found it unavoidable to raise our prices. The new prices will take effect from 1 October, and our revised price lists are attached.

We are sorry that these increases have been necessary, but can assure you that they will not amount to an average of more than about 5%. As general prices have risen by nearly 10% since our previous price list, we hope you will not feel that our own increases are unreasonable.

Yours sincerely

Letters of thanks

Business people have many opportunities for writing letters expressing appreciation and creating goodwill. Such letters of thanks can be as brief and as simple as you like, and they must express your appreciation with warmth and sincerity. The aim is to make the reader feel that you really mean what you say – and that you enjoy saying it.

In letters of appreciation, do not include specific sales matters or the reader may think that your thanks are merely an excuse for promoting business.

Email of thanks for a first order

Dear Mr Maxwell

You have already received our formal acknowledgement of your order number 456 dated 12 July. However, as this is your first order with us I felt I must write personally to say how pleased we were to receive it. We are glad of the opportunity to supply the goods you need.

I hope our handling of your order will lead to further business between us. We look forward to a happy and mutually beneficial association.

If you have any questions, please let me know.

Wendy Lim

Email of thanks for a large order

Dear Mrs Usher

I understand that you placed an unusually large order with us yesterday, and I want to thank you personally for your continued confidence in us.

Our happy working relationship for many years has always been valued and we shall do our best to maintain it.

All the best

Leslie Wong

Message of thanks for prompt settlement of accounts

Dear Mr Watts

I appreciate the promptness with which you have settled your accounts with us during the past year, especially as many of them have been for very large amounts.

This has been of great help to us at a time when we have been faced with heavy commitments connected with our business expansion.

I look forward to our great business relationship continuing in the future.

Best wishes

Daniel Tripp

Email of thanks for a service performed

Dear Mandy

Thank you for your email attaching the draft of our new product catalogue.

I am very grateful for the trouble you have taken to examine the draft and comment on it in such detail. Your suggestions and amendments will be very helpful.

I realise the value of time to a busy person like you, and this makes me all the more thankful for the time you have given so generously.

See you soon.

George Watson

Email of thanks for information received

Dear Mrs Webster

Thank you for your email attaching an article explaining the organisation and work of your local trade association.

I am very grateful for the interest you have shown in our proposal to include details of your association in the next issue of the *Trade Association Year Book*, and for your trouble in providing such an interesting account of your activities. This feature is sure to inspire and encourage associations in other areas.

All the best

Maria Jones

Letters of congratulation

One of the best ways to promote goodwill is to write a letter of congratulations. The occasion may be a promotion, a new appointment, receiving a special award, establishing a new business, examination success, even a marriage or a birthday. Your letter or note may be short and formal, or conversational and informal, depending on the circumstances and the relationship between you and the recipient.

Formal letter of congratulation on the award of a public honour

Letters of congratulation sent to mark the award of a public honour need only be short and formal. To show a sign of personal interest, when sending a formal letter it's a good idea to handwrite the salutation and complimentary close personally.

> I was delighted to learn that your work at the South Down College of Commerce has been recognised in the New Year Honours List. At a time when commercial education is so much in the public eye, it gives us all at the Ministry great pleasure to learn of your OBE.

Informal letter of congratulation on the award of a public honour

> On looking through the *Camford Times* this morning I came across your name in the New Years Honours List. I would like to add my congratulations to the many you will be receiving.
>
> The award will give much pleasure to a wide circle of people who know you and your work. Your services to local industry and commerce over many years have been quite outstanding. It is very gratifying to know that they have been so suitably rewarded.
>
> Very best wishes

Formal letter of congratulation on a promotion

> Dear Dr Roberts
>
> I would like to convey my warm congratulations on your appointment to the Board of Electrical Industries Ltd.
>
> My fellow directors and I are delighted that the many years of service you have given to your company have been rewarded in this way.
>
> We all join in sending you our very best wishes for the future.
>
> Yours sincerely

Letter of congratulation on employee's 10th anniversary

This month marks your 10th anniversary as a member of staff of SingComm Pte Ltd. We would like to take this opportunity to thank you for these past 10 years of fine workmanship and company loyalty.

We know that the growth and success of our company is largely dependent on having strong and capable staff members like you. We also recognise the contributions you make in helping us maintain the position we enjoy in the industry.

We are hoping that you will remain with us for many years to come and would like to offer our congratulations on this special anniversary.

 CAUTION Avoid waffle. Read through your message and tidy up anything that is repetitive or out of logical flow.

Letter acknowledging congratulations

Courtesy requires that letters of congratulation should be acknowledged. In most cases a short formal acknowledgement is all that is necessary.

This letter would be a suitable reply to the letter of congratulation on page 368. The writer very properly takes the opportunity to acknowledge her debt to colleagues who have supported her in her work.

Dear Mrs Fleming

Thank you for your letter conveying congratulations on the award of my OBE.

I am very happy that anything I have been able to do for commercial education in my limited field should have been rewarded by a public honour. At the same time I regard the award as being less of a tribute to me personally than to the work of my college as a whole – work in which I have always enjoyed the willing help and support of many colleagues.

Thank you again for your good wishes.

Yours sincerely

Letters of condolence and sympathy

Letters of condolence are not easy to write. There can be no set pattern to such letters since a lot depends on what kind of relationship the writer has with the recipient. As a general rule, such letters should usually be short and written with sincerity. To show special consideration, it's a good idea to handwrite letters of this kind.

Your letter should be written as soon as you learn the news. Express your sympathy in simple words that are warm and convincing, and say what you feel sincerely. Remember to write as if you are speaking.

Letter of condolence to a neighbour

Dear Mary

It was not until late last night that my wife and I learned of your husband's sudden and tragic death. This must have been a great shock to you. I want you to know how very sorry we both are, and we send you our sincere sympathy.

If there is any way in which we can be of any help, either now or later, please let us know. We shall be only too glad to do anything we can.

Yours sincerely

Letter of condolence to a customer

Dear Mr Kerr

I have just learned with deep sadness of the death of your wife.

There is not much one can say at a time like this, but all of us at Simpsons who have dealt with you would like to extend our sincere sympathy at your loss.

Please include us all among those who share your sorrow at this sad time.

Yours sincerely

Letter of condolence to a business associate

Dear Mrs Anderson

We were so sorry to read in *The Times* this morning that your Chairman has died, and I want to express our deep sympathy.

I had the privilege of knowing Sir James for many years and always regarded him as a personal friend. By his untimely passing, our industry has lost one of its best leaders. He will be greatly missed by all who knew him.

Please convey our sympathy to Lady Langley and her family.

Yours sincerely

Letter of condolence to an employee

Dear Maxine

I was very sorry to learn of your father's death. I remember your father very well from the years he served in our company's Accounts Department until his retirement two years ago. I well recall his love for his family and the great sense of pride with which he always spoke of his daughters.

He has been greatly missed at Wilson's since his retirement. We all join in expressing our sympathy to you and your family at this very sad time.

Yours sincerely

Letter of condolence to a friend

Dear Henry

I was so very sorry to hear the news of Margaret's passing.

Margaret was a very dear friend and we shall greatly miss her cheerful outlook on life, her generous nature and her warmth of feeling for anyone in need of help. Above all, we will miss her for her wonderful sense of fun.

Tom and I send you our love and our assurance of continued friendship, now and always. If there is any help you need at any time, just let us know.

We'll see you soon.

Yours

Alice

Letter of sympathy to a business associate

Dear Bill

When I called at your office yesterday I was very sorry to learn that you had been in a car accident on your way home from work recently. However, I was equally relieved to learn that you are making good progress and are likely to be back at work again in a few weeks.

I had a long talk with Susan Carson and was glad to learn of your rising export orders. I expect to be in Leicester again at the end of next month and shall take the opportunity to call on you.

Meanwhile, best wishes for a speedy recovery.

Yours sincerely

Acknowledgements of sympathy or condolence

You will naturally wish to acknowledge letters of the kind illustrated in this section. Such acknowledgements need only be short but they show that you are genuinely moved by the warm expressions of sympathy you have received.

Personal acknowledgement

Individual personalised acknowledgements should be made to relatives and close friends.

Dear Mrs Hughes

My mother and family join me in thanking you for your very kind letter on the occasion of my father's death.

We have all been greatly comforted by the kindness and sympathy of our relatives and friends. Both at home and in the hospital, where my father spent two weeks prior to his passing, the kindness and sympathy shown by everyone has been overwhelming.

Yours sincerely

Printed acknowledgement

When many letters of condolence have been received it will be sufficient to prepare a printed general acknowledgement.

Mr and Mrs Ashton and family thank you most sincerely for your generous expression of sympathy at their sad loss.

The kindness of so many friends and the many expressions of affection and esteem in which Margaret was held will always remain a proud and cherished memory.

97 Lake Rise
Romford
Essex
RM1 4EF

Checklist

■ Write and send letters promptly.

■ Use an appropriate tone.

■ Be sincere.

■ Use an informal style.

■ Use a personalised approach (handwritten if appropriate).

■ Keep them short and to the point.

■ Write as though you are having a conversation with the reader.

■ Ask yourself how you would feel on receiving such a letter.

23

Notices, advertisements, information sheets and handouts

Design skills

Notices, advertisements, information sheets and handouts are all documents that need an element of design. You need to encourage people to read them and take action on the content. Long, rambling paragraphs will not achieve this aim. It's good to have a great heading, great presentation and keep the information to a minimum.

When presenting creative documents like these, consider the AIDA principles:

Attention You must attract the reader's attention:
- use company logo?
- compose specific heading
- put special information in boxes/shaded section
- use sub-headings/numbered points/bullets

Interest Get the reader's interest by mentioning something that will appeal to them:
- be persuasive
- use simple language
- keep your sentences short

Desire Arouse the reader's desire to buy, to attend, to find out more or to contact the writer:
- make everything sound interesting
- point out the benefits

Action Make the reader want to do something after reading the document

Notices

Most organisations have notice boards at prominent places throughout the offices. These are used to bring special items to the attention of all staff. The information must be displayed attractively so that it gains attention and co-operation. Since the introduction of the intranet, this is also a place where notices are posted to bring items to everyone's attention. Intranets are a great way to facilitate communication between people or workgroups.

Notices may be posted about:

- new procedures
- social events
- advertisements for internal appointments
- reports on matters of interest
- reminders of company procedures.

Notice

STAFF CAR PARK

CLOSED

URGENT REPAIR WORK
will be made to the staff car park
from 8–10 August

Please note that the car park will be closed during this period.

All staff must:

- avoid using their cars where possible
- share cars and give each other lifts to and from work

If you really have to use your car to come to work, please use the customers' car park. However, please remember that space is limited.

David Yap
Manager

DF/ST
25 July 201-

Notice

<table>
<tr><td>Headline to draw attention ——</td><td align="center">CAN YOU ACT?
(OR WOULD YOU LIKE TO TRY?)</td></tr>
<tr><td>Make it sound interesting ——</td><td>This year Global Communications celebrates its 10th year of providing quality telecommunications equipment. To mark this special occasion we are holding a 10th Anniversary Celebration at the Regal Prince Hotel in July. Special guests, directors and staff are to be invited.</td></tr>
<tr><td>Give full details ——</td><td>A special 20-minute sketch has been written and we are now looking for aspiring actors and actresses to perform the sketch.</td></tr>
<tr><td>Use another heading if relevant ——</td><td align="center">INTERESTED?</td></tr>
<tr><td></td><td>If you would like to attend an audition for a part in this sketch please come along and find out more.</td></tr>
<tr><td>Use attractive layout to draw attention to special points ——</td><td>When?   Monday 15 March
Where?   Training Room, Global House
Time?   6.00–8.00 pm

If you will be coming along to the auditions please call Mandy Jones, Marketing Department, to put your name down.</td></tr>
<tr><td>Include name/title at foot of notices ——</td><td align="center">James Porter
Marketing Manager</td></tr>
<tr><td>Always date notices ——</td><td>JP/ST
2 March 201-</td></tr>
</table>

CAUTION Avoid using too many fonts or fancy bullets in notices. This will just detract from the message.

Advertisements

Most organisations advertise in newspapers, magazines or trade journals to reach out to a wide and sometimes specific market. Advertisements can also be posted on the company's intranet. Advertisements may be placed to:

- advertise vacant posts
- promote products or services
- announce special events or functions
- publicise changes in the organisation.

Job advertisement

Company logo

Headline attracts attention

Job title

Information about the company

Use bullet points

State what you want from applicants

Emphasise compensation package

GLOBAL COMMUNICATIONS

Together We Excel

QUALITY ENGINEER

Here's your chance to join a dynamic and rapidly expanding global company. Global Commmunications is a global leader in the world of telecommunications. We are looking for dynamic, self-motivated team players to join us.

Responsibilities
- Ensure Global's major accounts continue to benefit from customer service and support.
- Support major accounts on technical and quality issues.
- Organise pre-sales initiatives on new products.

Pre-requisites
- Experience in telecommunications industry.
- ISO 9000 knowledge.
- Good people skills.

We offer a 5-day week and attractive compensation package commensurate with experience and qualifications.

State if applications are — to be handwritten

Name/title/company — name and address

Closing date for — applications

Please send a handwritten application letter with full résumé to:

Mr William Wright
Human Relations Manager
Global Communications plc
14 Western Avenue
Highbury
Sheffield S26 2ES

Closing date for applications: 12 July 201-

Notice or advertisement announcing a job

GLOBAL COMMUNICATIONS
require a
TEAM LEADER – CUSTOMER SERVICES

Do you have what it takes to build a successful team? We are looking for someone who can coach, support and develop individuals to improve team performance.

The successful applicant will have:

- excellent supervisory skills
- proven skills in directing and controlling projects
- good organisational skills
- effective decision-making skills
- first-class leadership skills

Global Communications provide great working conditions:

- flexible working hours
- excellent training and development
- private medical insurance (after one year)
- 20 days' holiday allowance, increasing with length of service
- incentive schemes
- career opportunities.

IF YOU THINK YOU HAVE WHAT IT TAKES
please apply ONLINE ONLY at
www.global.com.sg

TIP Make your advertisement eye-catching so that it stands out clearly from the rest.

Leaflets

Many types of leaflets and information sheets are produced in business today. They are useful to convey information to a large number of people. Leaflets and information sheets may be handed out in the street or at the entrance to shopping centres, left in public areas for people to pick up, at museums and exhibitions, at department stores and shops, or even delivered to your home inside newspapers and magazines.

Information sheets may give details about any number of things, including:

- goods or services
- special promotions
- special events or functions
- guidelines and information

Leaflets can be a great marketing technique, but the success will depend largely on the leaflet design. A leaflet must grab the recipient's attention, make them interested in the product or service, and finally induce some action, whether that's phoning a number, visiting a website, or placing an order.

Here is a step-by-step guide for creating an effective leaflet:

1 Make it look professional

Professional leaflets make a big difference. Anyone can write a few lines of text on their home computer, print it out in black and white, and then deliver it to households. However, unprofessional leaflets will reflect badly on your organisation, and will possibly do you harm than good. Let's face it, why should people feel confident ordering from you if you can't spend a little money promoting your business?

2 Use an attention-grabbing headline

Many people pick up leaflets that have been delivered to their home and head straight for the bin. In the time it takes for them to walk there, your leaflet needs to grab their attention. This is why you need an excellent headline. Don't use your business name or product category as the headline. Instead, create a headline that initiates interest, such as 'What happens when your washing machine breaks down?' Find out what's important to your customers and incorporate that into your headline.

3 Keep your sentences short

People rarely read long sentences on leaflets so don't use them. Bullet points are a much better idea, and they help you get straight to the point.

4 Include 'power words'

Some words are known for being good at attracting attention. Such words include: avoid, bargain, bonus, earn, easy, enjoy, exciting, exclusive, discover, fast, free, how to, learn, money, mystery, new, profit, save, special, win. Try to naturally incorporate some of these words into your text.

5 Use images

The saying goes that a picture says a thousand words, so think about using images. Make sure it's relevant and reinforces the brand image that you're trying to portray or what you are saying.

6 Get rid of clutter

You don't have to fill every available space on the leaflet. A little blank space looks easier on the eye and makes it easier to read.

7 Include a call to action

Tell people what to do next, but not in a passive way. 'For more information, call . . .' is a passive call to action. It doesn't motivate action. This sort of thing would be much more action-oriented:

- Buy before (date) and receive this gift
- Book before (date) and get 10% off
- Buy now – 10% off for the first 50 customers

Leaflet

MISUSE OF ANTIBIOTICS CAN PUT YOU AND OTHERS AT RISK

Antibiotics can be lifesavers, but misuse has increased drug-resistant germs. See how this affects you and what you can do to help prevent antibiotic resistance.

If you think antibiotic resistance is not a problem or doesn't affect you, think again. A prominent example of the dangers of antibiotic resistance is the spread of MRSA, or Methicillin-Resistant Staphylococcus Aureus. MRSA was once a concern only for people in hospital, but a newer form of MRSA is causing infections in healthy people in most communities.

Antibiotic resistance occurs when antibiotics no longer work against disease-causing bacteria. These infections are difficult to treat and can mean longer lasting illnesses, more doctor visits or extended hospital stays, and the need for more expensive and toxic medications. Some resistant infections can even cause death.

Did you know?
- Antibiotics have no effect on viral infections (eg colds, flu and most sore throats).
- Viral infections are more common than bacterial infections.
- Inappropriate use of antibiotics can help to develop resistant bacteria.
- If you take antibiotics too often, or when not really needed, they may not work when you really need them.
- Some antibiotics have harmful side effects, such as diarrhoea and allergic reactions.
- Antibiotics do not just attack the infection they are prescribed for. They also kill useful bacteria, which protect you against other infections, such as thrush.
- Illnesses are already cropping up that are resistant to even the strongest antibiotics.

> When you really need to take antibiotics, ask your doctor also to prescribe some probiotics. These will help to counter-balance the effects of antibiotics by putting back the good bacteria into your body.

When should I use antibiotics?
Antibiotics are effective against bacterial infections, certain fungal infections and some kinds of parasites.

This chart shows common illnesses and whether they are caused by bacteria or viruses. Taking an antibiotic when you have a viral infection will not make you better, and can contribute to antibiotic resistance.

Bacterial infections	Viral infections
■ Some ear infections	■ Most ear infections
■ Severe sinus infections	■ Colds
■ Strep throat	■ Influenza (Flu)
■ Urinary tract infections	■ Most coughs
■ Many wound and skin infections	■ Most sore throats
	■ Bronchitis
	■ Stomach flu (viral gastroenteritis)

Consequences of antibiotic misuse
- If antibiotics are used too often for things they cannot treat, they become less effective against the bacteria they are intended to treat.
- When bacteria become resistant to first line treatments, the risk of complications and death is increased.
- Not taking antibiotics exactly as prescribed can lead to problems. For example, if you take an antibiotic for only a few days, instead of the full course, the antibiotic may wipe out some but not all of the bacteria. The surviving bacteria become more resistant and can be spread to other people.
- Thousands of people die each year from antibiotic-resistant infections that they contracted in hospital.

Handouts

Effective handouts are an integral part of most technical presentations, so their design, use and distribution require careful planning. To make your handouts more effective, follow these simple rules:

1 Keep it simple

Avoid unnecessary details, focus on key words and concepts from your presentation. You don't want people to focus on your handout. You want them to listen to you and glance at your handout as you speak. Use some graphics or illustrations to help make your point, and leave some room for notes where appropriate too.

2 Make sure the notes follow your presentation

There's nothing worse than participants flipping all over the place on handouts because they don't know where the trainer is. Make sure your points correspond to the points you are making in your presentation. If you jump around, your audience will have to spend time figuring out where you are.

3 Make sure it looks good

Use plenty of white space, small chunks instead of lengthy paragraphs, avoid anything too fancy, and keep to one font or style.

Handout

Use a clear heading ————————— **PASSIVE vs ACTIVE VOICE**

Which is more effective? Why?

#1 Your order was received by us today. The goods you requested have been despatched by courier. You should receive them within 48 hours. Should they not have been received by you tomorrow our despatch department should be contacted as soon as possible.

#2 Thank you for your order number HT121 dated 21 June. Our special courier service will deliver these goods within 48 hours. If you have not received them by Friday morning please contact me on 254 8777.

Leave some blanks so that you can elicit details from students

Why NOT *passive* voice?	**Why USE *active* voice?**
■ It makes your writing vague.	■
■ It denies responsibility.	■
■ It creates a distance between you and your reader.	

Allow some practice ——— ✍ **YOUR TURN**

Rewrite these sentences using ACTIVE voice.

1. The seminar will be conducted by Robert Sim.
2. The leak was fixed by the plumber.
3. Your co-operation will be greatly appreciated.
4. Arrangements have been made for the conference to be held at the Hilton Hotel.
5. The investigation has been concluded by our client, and the paperwork has been signed.
6. The design of our new systems was simplified by the use of hydraulics.

Use sub-headings ——— **Is *passive voice* ever appropriate?**

✓ **In minutes of meetings**	Mrs Jones reported that the photocopier had broken down for the third time in a month.
✓ **When tact is important**	A serious mistake was made.

TIP Use sub-headings, numbers and bullets to make your handouts attractive and easy to follow.

 Checklist

■ Include a company logo, prominently displayed, if appropriate.

■ Compose a catchy headline to give the gist of the contents.

■ Use spacing carefully to give prominence to special items.

■ Use sub-headings and shaded sections to attract attention.

■ Use numbered points and bullets to categorise information.

■ Ensure your display is eye-catching, attractive and effective.

■ Use straightforward, simple language and short sentences.

■ Use persuasive and convincing writing skills to make everything sound useful, exciting, interesting or beneficial.

■ Aim for your document to stand out when alongside many others.

■ State the action that you want the reader to take and if necessary include name, telephone number, email address, etc.

Circulars

24

ircular letters are used to send the same information to a number of people. They are extensively used in sales campaigns and for announcing important developments in business, such as extensions, reorganisations, changes of address, etc. When writing circulars, here are some guidelines to remember:

1 **Know your audience**. Readership for circular letters can be quite diverse, so it's hard to guess your audience's level of prior knowledge or familiarity with the content. However, if you aim to make it very generic and broad-based, your letter will be understandable and useful to most people.

2 **Use appropriate tone**. Your tone and style should be relevant for the kind of communication (eg internal or external). For example, the tone would be quite firm in a circular going out to all employees telling them the company's phone bills are enormously high and asking them to cut down unnecessary personal calls. However, this tone would not be appropriate to use in a letter going out to clients.

3 **Share only authorised information**. Since circular letters are intended for a large audience, you must not give away any confidential information.

4 **Be concise**. No one wants to read a long-winded message with lots of padding. Get straight to the point and keep it brief.

5 **Use a personal touch**. Create the impression of personal interest by using language like *you*, never *our* customers, you all, all staff, our clients, everyone.

Instead of	Say
Our customers will appreciate . . .	You will appreciate . . .
We are pleased to inform all our clients . . .	We are pleased to inform you . . .
Everyone will be interested to learn . . .	You will be interested to learn . . .
Anyone visiting our new showroom will see . . .	If you visit our new showroom you will see . . .

Circulars announcing changes in business organisation

Changes in a firm's business arrangements may be announced by circular letters such as those which follow. Where the salutation has been left blank, it has been presumed that the letter would be word processed and individual names, addresses and salutations would be merged to add a personal touch.

Transfer of business

Dear Customer

CHANGE OF COMPANY NAME/TRANSFER OF BUSINESS

In November 201- Merlion Communications was acquired by SingComm Pte Ltd. As a result of this acquisition and renaming the company, we are amending the registered name for direct debit processing.

This change will not affect the service you receive in any way, except that future direct debits will be collected by SingComm Pte Ltd instead of Merlion Communications with immediate effect. The only change you will notice is the different name on your bank statement.

You need not take any action. Details of the change have been sent to your bank. Your rights under the direct debit guarantee are not affected, as detailed on the attached guarantee.

Thank you for your continued support.

Yours sincerely

Change of company name

Dear

We are pleased to announce that further to the 100% acquisition by FGB Insurance (Asia Pacific) Holdings Limited, Ruben Insurance Pte Ltd has been renamed FGB Insurance (Singapore) Pte Ltd. General insurance operations will start using this new name from 2 August 201-.

You will continue to enjoy the same high level of service that you have previously received from Ruben Insurance Pte Ltd. You will also see additional benefits arising from the wide-ranging expertise, products and services of the FGB Group, as well as the strong financial standing that FGB brings to our 34 million customers all over the world.

With effect from 2 August 201- we will be relocating to this new address:

45 Robinson Road, #02-04 Wisma Supreme, Singapore 234381

Our new telephone number will be +65 63453456.

Please visit our website at www.fgbins.com.sg for the latest information.

Your current insurance policy remains legally valid and we will honour all our obligations and liabilities under documents bearing our former corporate name.

If you have any questions at any time please call us on 63453456.

We thank you for your support and look forward to being of great service to you.

Yours sincerely

Notification about new association

Dear Client

It gives us great pleasure to announce that on 1 April we have entered into a close association with Garner Accountancy Co Ltd of 22 High Street, Cheltenham.

We have formed a new company that will operate as Garner and Barret Accounting Co Ltd, and as a result we will be moving to bigger premises at:

21 Hillington Rise
Sheffield
S24 5EJ

Telephone: 0114 2874722
Fax: 0114 2874768
Website: **www.garnerbarret.co.uk**

This association provides us with a much bigger base that will enable us to offer improved services to our customers. We will of course ensure that we retain the close personal contact and interest in our clients' affairs.

We also take this opportunity to announce that Mr Robin Wilson, who is already known to many clients, will become a partner in the new company with effect from the same date.

Yours sincerely

TIP In direct mail letters or emails, address the reader directly by using 'you' to get their attention and make it more personal.

Opening of a new store

Dear

BEST SUPERSTORE OPENS IN BEDFORD – 12 JULY 201-

Have you seen the great news in the national newspapers recently? Best International is opening a chain of furniture superstores throughout the UK. The first one will open in Bedford on Monday 12 July 201-.

The first 50 customers who come through our doors on that day will receive special discounts. Opening times are:

0900–2400 Monday to Saturday

0900–1700 Sunday

You will find a variety of modern kitchens, bathrooms, dining rooms and lounges on display. A full planning service is available so you can leave it to the experts to design exactly what you want. Each department in the Superstore is supervised by friendly, qualified staff.

The store will be of particular interest to you if you're a DIY enthusiast. You will find everything you may need – paints, wall coverings, tiles, carpets, and so much more. We will deliver free of charge any orders over £100 – for smaller orders there will be a minimal charge. Credit facilities are available at low interest rates.

Our car park has spaces for 400 cars, but if you prefer to take the bus, number 214 stops right outside the Best Superstore.

Don't miss our grand opening on Monday 12 July. **Remember there's a special discount waiting for you if you are among the first 50 customers.**

See you at the BEST Superstore!

Expansion of existing business

Dear Customer

To meet the growing demand for a hardware and general store in this area, we have decided to extend our business by opening a new Hardware Department.

This new department will carry an extensive range of domestic goods and hardware at prices that compare favourably with those charged by other suppliers.

We hope you will give us the opportunity to demonstrate our new merchandise to you, so we are arranging a special window display during week beginning 24 June. The official opening of our new Hardware Department will be on the following Monday 1 July.

We look forward to welcoming you during opening week. We willl show you that the reputation enjoyed by our other departments for giving sound value for money will apply equally to this new department.

Yours sincerely

Opening of a new business

Dear Householder

We are pleased to announce the opening of our new retail grocery store, **Great Value**, on Monday 1 September.

Mrs Victoria Chadwick has been appointed Manager. She has 15 years' experience of the trade and we are sure that the goods supplied will be reasonably priced and great quality.

Our new store will open at 0800 on Monday 1 September. As a special celebration offer, we will be giving you a discount of 10% on all purchases made if you're among the first 50 customers.

We hope we can look forward to seeing you at **Great Value**.

Yours sincerely

CAUTION Put yourself in the reader's position. Is anything unclear in your message? Will they have to ask questions? Are misunderstandings possible? If so, revise.

Establishment of a new branch

Dear

Our sales in the Kingdom of Jordan have increased so much recently that we have decided to open a branch in Amman. Mr Faisal Shamlan has been appointed as Manager.

We hope we have provided you with an efficient service in the past, and we are very excited to continue working with you now that we have a presence in Amman. Our new branch in your country will mean we will be able to deal with your enquiries and orders much more promptly.

This new branch will open on 1 May, and from that date all enquiries should be directed to:

Mr Faisal Shamlan
Manager
Tyler & Co Ltd
18 Hussein Avenue
Amman
Tel: (00962)6-212421
Fax: (00962)6-212422
Email: faisal.shamlan@Tyler.com

Thank you for your valued custom in the past, and we hope these new arrangements will lead to an even better business relationship with you.

Yours sincerely

Removal to new premises

Dear

The steady growth of our business has made necessary an early move to new and larger premises. We have been fortunate in being able to move to a particularly good site on the new industrial estate at Chorley, so from 1 July our address and contact details will be:

Unit 15
Chorley Industrial Estate
Grange Road
Chorley
Lincs CH2 4TH

Telephone 456453
Fax 456324

This new site is served by excellent transport facilities, both road and rail, enabling deliveries to be made promptly. It also allows us to use much better methods of production, which will increase output and improve the quality of our goods even further.

Thank you for your valued custom in the past, and we confidently expect to be able to offer you improvements in service when the new factory moves into full production.

Yours sincerely

TIP The four-point plan is always a useful way to structure your documents logically. Find out more in Chapter 4.

Death of a colleague

Dear

It is with much sadness that I have to tell you of the sudden death of our Marketing Director, Michael Spencer. Michael had been with this company for 10 years and he made an enormous contribution to the development of the business. He will be greatly missed by all his colleagues.

It's important that we continue to provide efficient service to you. Please contact me personally with any matters that Michael would normally deal with.

Yours sincerely

Circulars announcing changes in company representatives

When a change takes place in the senior members of a company or membership of a partnership, it's important to tell suppliers and customers as soon as possible. For retiring partners, this is particularly important since they remain liable not only for debts contracted by the firm during membership but also for debts contracted with old creditors in retirement.

Retirement of a partner

Dear

I am sorry to inform you that our senior partner, Mr Harold West, has decided to retire on 31 May due to recent extended ill-health.

The withdrawal of Mr West's capital will be made good by contributions from the remaining partners, and the value of the firm's capital will therefore remain unchanged. We will continue to trade under the name of West, Webb & Co, and there will be no change in policy.

We hope that the confidence you have shown in our company in the past will continue, and that we may rely on your support in the future.

Yours sincerely

 CAUTION Take care with commas. They are often placed wrongly. In the first paragraph here, you'll notice a comma correctly placed before and after Mr Harold West. Find out more about punctuation in Chapter 2.

Appointment of a new partner

Dear

A large increase in the volume of our business has made necessary an increase in the membership of this company. It is with pleasure that we announce the appointment of Mrs Briony Kisby as partner.

Mrs Kisby has been our Head Buyer for the past 10 years and is well acquainted with every aspect of our policy. Her expertise and experience will continue to be of great value to the company.

There will be no change to our firm's name of Taylor, Hyde & Co.

We look forward to continuing our mutually beneficial business relationship with you.

Yours sincerely

Conversion of partnership to private company

Dear

The need for additional capital to finance the considerable growth in the volume of our trade has made it necessary to reorganise our business as a private company. The new company has been registered with limited liability in the name Barlow & Hoole Limited.

This change is in name only and the nature of our business will remain exactly as before. There will be no change in business policy.

Our personal relationship that we have developed in the past will be maintained. We shall continue to do our utmost to ensure that you are completely satisfied with the way in which we handle your future orders.

Yours sincerely

Appointment of new Managing Director

Dear

NEW MANAGING DIRECTOR

We are pleased to announce the appointment of Richard Wilson as Managing Director with effect from 2 September 201-. His appointment follows the early retirement in July of Francis Billington due to ill-health.

Richard is already known to many of you through his position as Marketing Director. He has 12 years' experience with Yangon Electrics, and he is looking forward to taking over this more challenging role in the company.

We are happy to assure you that we shall continue to provide the high-quality service for which we are proud to enjoy such a good reputation.

If you have any urgent queries, please contact me personally.

Yours sincerely

 CAUTION In the last paragraph, note the use of 'please' and not 'kindly'. Use modern language in your messages. Find out more in Chapter 3.

Dismissal of firm's representative

Dear

Miss Rona Smart, who has been our representative in North-West England for the past seven years, has left our service. Therefore, she no longer has authority to take orders or to collect accounts on our behalf.

Mrs Tracie Coole has been appointed to this position immediately. Mrs Coole has had control of our sales section for many years and she is thoroughly familiar with the needs of customers in your area. She intends to call on you sometime this month to introduce herself and to bring samples of our new spring fabrics.

We look forward to continuing our business relationship with you.

Yours sincerely

Note that if the representative left of her own free will and was a valued member of staff, the first paragraph of the above letter would be more suitably expressed as follows:

We are sorry to inform you that Miss Rona Smart, who has been our representative for the past seven years, has decided to leave us to take up another appointment.

Appointment of new representative

Dear

Mr Samuel Goodier, who has been calling on you regularly for the past six years, has now joined our firm as junior partner. His many friends will undoubtedly be sorry that they will see him much less frequently, and we can assure you that the feeling is mutual.

Mr Goodier hopes to keep in touch with you and other customers by occasional visits to his former territory.

Mr Lionel Tufnell has been appointed to represent us in the South West. Mr Goodier will introduce him to you when he makes his last regular call on you next week.

Mr Tufnell has worked closely with Mr Goodier in the past and he will continue to do so in the future. Mr Goodier will continue to offer help and advice in matters affecting you and other customers in the South West. His intimate knowledge of your requirements will be of great benefit to Mr Tufnell in his new responsibilities.

Our business relations with you have always been excellent, and we believe we have served you well. We hope you will extend to our new representative the courtesy and friendliness you have always shown to Mr Goodier.

Yours sincerely

 TIP Be clear and concise in your writing. Using short words and simple expressions will help you.

General circulars to clients

Notification of price increase

Dear

I am sorry to inform you that, due to an unexpected price increase from our manufacturers in Europe, we have no option but to raise the prices of all our imported shoes by 4% from 6 October 201-.

Orders received before this date will be invoiced at the old price levels.

We sincerely regret the need for these increased prices. However we know you will understand that this increase is beyond our control.

We look forward to a continuing association with you, and can assure you of our continued commitment to good-quality products and service.

Yours sincerely

Invitation to special function

Dear

10TH ANNIVERSARY CELEBRATION

Omega International is commemorating its 10th year of providing quality communications equipment. We are planning to hold a special celebration in August.

As one of our major clients, we are pleased to invite you to join over 100 of our management and staff to attend this celebration. Details of the function are:

Where?	Orchid Suite, Merlion Hotel, Orchard Road
When?	Friday 27 August 201-
What time?	Cocktails 6.30 pm
	Dinner 7.30 pm

There will be many highlights during this special evening, including speeches and special awards to clients and employees, plus lucky draw prizes and a cabaret act.

Please let us know whether you will be able to attend by returning the enclosed reply form before 31 July or by telephoning Iris Tan on 64545432.

We do hope you will join us to help make this evening a success.

Yours sincerely

TIP This is a modern way of displaying the details of an event. You could also use the side headings 'Venue, Date, Time'. Refer to Chapter 16 to learn more about using bullets and lists in your messages.

Survey of customer attitudes

Dear

Mansor Communications are committed to providing quality service, and as such we like to keep in touch with customer needs and views on the products that we sell.

To maintain our high standard of quality products and services to you, I hope you will take a few moments to complete the enclosed questionnaire. In appreciation of your trouble, I shall be pleased to send you one of our superb Mansor Pens on receipt of your completed questionnaire.

I look forward to receiving your reply, and can assure you of our continued good service to you in the future.

Yours sincerely

Internal circulars to staff

Many circulars are written to staff regarding various matters concerning the general running of the business, safety and security, administrative matters and many other things. It's most likely that these circulars will be sent via email these days.

Announcement about new working hours

Subject: New working hours

Dear colleagues

With effect from 1 September 201- working hours will be changed to 0930 to 1730 Monday to Friday instead of the present working hours of 0900 to 1700.

If you anticipate experiencing any difficulties please let your manager know before 14 August.

Notice about new car park

Subject: New company car park

Dear all

As you know, some old buildings on our site have been demolished. A piece of land in this area has been cleared so that it may be used as a car park.

The new car park should be ready for use by 28 October. It will be available between 0730 and 1830 hours Monday to Friday. The company takes no responsibility for loss or damage to vehicles or contents while in the car park.

If you wish to use the car park, please obtain an agreement form from Mr John Smithson, Security Officer. You will need to complete this form and return it to him before using the car park.

Many thanks

Information about store discount

Subject: Staff discounts at Quantum Stores

Dear all

We have recently reached an agreement with Quantum Stores that allows all our employees to receive 10% discount on any goods that are not already reduced in price. A discount of 2.5% will be given on reduced price or sale goods.

To claim the discount you must show your Omega identification badge at the checkout. These discounts will take effect from 1 September 201-.

Security information to Heads of Department

Subject: Security

Dear Department Managers

In view of recent bomb threats received by several competitors, please brief your staff on these points of security.

1. All employees must wear a name badge at all times.
2. All areas must be kept as clean and tidy as possible. This will reduce potential areas where bombs may be hidden.
3. Do not tamper with or move any suspicious object. Department Managers should be informed and the police notified.
4. Evacuation should follow established fire drills.

All incidents must be taken seriously and a detailed report submitted to me.

Please stress to all your staff that they have an important part to play in maintaining a high level of security in all areas at all times.

Thanks for your help.

Letter regarding outstanding holiday entitlement

In the past it has been a policy of the company that all staff must take their holiday entitlement within one calendar year. Any holiday entitlement not taken before 31 December each year has been forfeited.

It has now been decided to amend this rule so that we can allow our staff more flexibility.

With immediate effect, anyone who has up to five days' holiday entitlement outstanding at 31 December may carry this over until 31 March the following year. Any days that have not been used by 31 March will be forfeited. Unused holiday entitlement may not be converted to pay in lieu.

Your manager/supervisor is still required to approve your staff leave. This will take into account the business and operational needs of the department and especially clashes with other staff.

If you have any questions about this new policy, please speak to your manager/supervisor, or call the Human Resource Department on extension 456.

Reminder about health and safety policy

In view of the recent unfortunate accident involving a visitor to our premises, I would like to remind you about our health and safety policy.

For security reasons, if you see someone you do not recognise wandering around the building unaccompanied, do not be afraid to ask questions. Ask politely why he or she is here, which member of staff they are visiting, and if that person knows the visitor has arrived.

All Merlion Tools employees have a responsibility to take reasonable care of themselves and others and to ensure a healthy and safe workplace. If you notice any hazard or potential hazard, please inform the Health and Safety Manager, Michael Wilson, who will investigate the issue immediately.

I am attaching the company's health and safety policy. Please take a few minutes to read it through to remind yourself of the main points.

Thank you for your help in ensuring the health and safety of all employees and visitors to Merlion Tools.

 TIP Read sentences out loud to hear the pace and rhythm. This will help you get the punctuation in the right place.

Circulars with reply forms

A tear-off slip is often used when a reply is required from people to whom a circular is sent. Alternatively a separate reply form may be used. The important points to remember with such reply sections are:

- begin with 'Please return by ___ to ___'. This is a safeguard in case someone separates the tear-off portion or reply form from the main letter.
- use double spacing for the portion which will be completed.
- leave sufficient space for completion after each question/heading.
- use continuous dots where answers are required.

Invitation to function (with tear-off slip)

Dear

For internal use, just include name/title; the full address is not required — 10TH ANNIVERSARY CELEBRATION

Omega International is celebrating its 10th year of providing quality communications equipment. Approximately 50 representatives from Omega clients are expected to attend a special 10th Anniversary Celebration on Friday 29 October 201-.

Keep it simple and precise — As you have been with Omega for at least 5 years we are pleased to invite you to attend this 10th Anniversary Celebration. This special

Use double spacing — function will take place from 1800 to 2300 hours at The Mandarin Suite, Oriental Hotel, West Street, London. Cocktails and a buffet dinner will be provided.

Please let me know whether you will be attending by returning the tear-off portion before 31 August.

I hope you will be able to join us.

....

Please return to Mrs Judy Brown, Administration Manager, before 31 August.

I shall/shall not* be attending the 10th Anniversary Celebration on Friday 29 October.

Name ....

Designation/Department ....

Signature .... Date ....

Remember footnote where appropriate — * Please delete as applicable.

Reply form

Here is a reply form to accompany a letter sent to clients of a training organisation. Clients were asked to specify whether or not they would like to attend a one-day management conference, when accommodation would be required, and enclose a cheque to cover the cost.

REPLY FORM

Please complete and return by 15 February 201- to

Give full address when used externally; don't forget reply date —— Mr Edward Teoh
Personnel Manager
Professional Training Pte Ltd
126 Buona Vista Boulevard
Kuala Lumpur
Malaysia

Use same heading as on covering document —— ONE-DAY MANAGEMENT CONFERENCE
SATURDAY 3 APRIL 201-

Use numbered points if appropriate —— 1. I wish/do not wish* to attend this conference.

Use the personal term 'I wish, require', etc —— 2. I require accommodation on (please tick):
    ☐ Friday 2 April

Use option/boxes where appropriate ——   ☐ Saturday 3 April

3. My cheque for M$400 is attached (made payable to Professional Training Pte Ltd)

Signature .... Date ....

Name (in caps) ....

Choose appropriate details at the foot —— Title ....

Company ....

Address ....

Leave sufficient space —— ....

.... Post code ....

Telephone .... Fax ....

Don't forget footnote —— * Please delete as necessary.

Checklist for circular letters

■ Personalise circular letters by merging individual relevant details.

■ Address the recipient by name if you know it, otherwise for emails you may use Dear colleagues, Hi staff, Hi team, Hi all, etc.

■ Make it personal – use 'you' instead of 'all customers', 'everyone'. Write as though you are speaking to one person, because only one person will read it each time.

 Checklist for tear-off slip/reply form

■ State reply date.

■ Mention to whom the form should be returned.

■ Internal forms – name/title only.

■ External forms – name/title/company name and address.

■ Use same heading as covering document.

■ Use double spacing.

■ Use personal terms – I wish, I shall/shall not.

■ Use options/boxes where appropriate.

■ Leave sufficient space for completion.

■ Ensure the form contains everything you need to know.

25

Sales letters and voluntary offers

Sales letters

A sales letter is the most selective of all forms of advertising. Unlike press and poster advertising, a sales letter aims to sell particular kinds of goods or services to selected types of customers. The purpose of a sales letter is to persuade readers that they need what you are trying to sell, and influence them to buy it.

In brief, you take something attractive and make it seem necessary, or you take something necessary and make it seem attractive.

 TIP Although sales letters will be circulated to many people, they must be written with a personal touch. Use 'you' rather than 'all of you'.

The weakest link in your sales letters

Here is my compilation of some of the weakest links that need attention in your sales letters.

1 Weak headline

The headline is probably the most important part of your sales letter. It is your first opportunity to capture your reader's attention and make him/her interested in what you have to offer. Powerful headlines in your sales letters can increase your profits substantially.

2 Weak writing

If your customers fall asleep while reading your sales letters, something is wrong. Your letters should hold the reader's attention from start to finish. They should be worded in a way that sounds interesting and exciting, with a sharp and snappy writing style.

3 Weak range of purchasing options

An easy way to attract sales is by offering customers more ways to pay. Most customers will want to use credit cards for convenience. However, you must also attract customers who prefer other payment methods, such as cheque, debit card, online purchasing, etc.

4 Weak selling proposition

What makes your product or service different from your competitors'? Lower prices, higher quality, faster delivery, a wider selection of products? Customers want to see a logical, realistic reason for doing business with you. You need to home in on a unique selling proposition – something you can deliver that your competition cannot. Be sure to mention this in all your advertisements and sales letters.

5 Weak or no endorsements

Endorsements can really boost your sales. A good endorsement is equal to a positive word-of-mouth referral. Potential customers trust what satisfied customers say about your products or services. When you receive favourable comments from customers, always ask them for permission to quote them and use their comments as endorsements in your sales letters.

6 Weak guarantee

Short guarantees are not popular with customers because they don't offer much time to find out whether your product or service works for them. Longer guarantees generally boost sales and reduce the number of returns.

7 Weak follow-up plan

The majority of visitors to your website will probably not buy anything from you the first time around. You must design an effective follow-up plan so that you maximise potential sales. One suggestion here is to offer a free sample of something in exchange for contact information.

Successful sales letters

A good sales letter must be structured to follow these general AIDA principles:

Attention	You must attract the reader's attention with a striking heading, perhaps a question or challenging statement.
Interest	Mention something that will interest the reader. Appeal to a particular motive, eg health, image, economy, fear.
Conviction	You must also show conviction in this section so the reader has confidence in you. Tell them how many years you've been in business, or what guarantees you can give.
Desire	Arouse the reader's desire. Describe the benefits rather than features. Emphasise what's unique, topical or new. Give guarantees or evidence of testing. Perhaps you can give an incentive to reply early?
Action	Make the audience want to buy, to find out more, to make a call. Make it clear what the reader must do next.

Create attention

In your opening paragraph you must create attention and encourage the reader to take notice of what you have to say. If you're not careful in this opening paragraph your letter could end up in the waste-paper bin without being read. It's often a good idea to start with a question, an instruction, a challenge, or a quotation. Here are some examples:

An appeal to self-esteem

Are you nervous when asked to propose a vote of thanks, to take the chair at a meeting, or to make a speech? If so this letter has been written specially for you!

An appeal to economy

Would you like to cut your domestic fuel costs by 20 per cent? If your answer is 'yes', read on . . .

An appeal to health

> 'The common cold', says Dr James Carter, 'probably causes more lost time at work in a year than all other illnesses put together.'

An appeal to fear

> More than 50 per cent of people have eye trouble and in the past year no fewer than 16,000 people in Britain have lost their sight. Are your eyes in danger?

Create desire

Now that you have the reader's interest in the opening paragraph, you must create a desire for the product or service you are selling. To do this, point out the benefits to the readers and how it will affect them.

If the letter is sent to a person who knows nothing about the product, you must describe it and give a clear picture of what it is and what it can do. Study the product and then select those features which make it superior to others of its kind. Stress the features from the reader's point of view.

To claim that a particular CD system is 'the best on the market' or 'the latest in electronic technology' is of little use. Instead, stress such points as quality of the materials used and the special technology that make the equipment more convenient or efficient than its rivals, as in the following description:

In this description, note that the final statement, 'as finely finished as a Rolls-Royce', equates the product with one which is well known and high-end. This creates a picture of a reasonably priced yet superb product

> We are proud of our standing as a modern design icon, and are constantly upgrading with the latest technological developments. This deeply functional music system immediately catches the eye with its bold combination of strong lines and random colours of the discs. When users load their own CDs, each player becomes unique, the perfect combination of human and machine. A true kinetic masterpiece, this system is an advanced piece of mechanical engineering, as finely finished as a Rolls-Royce.

Carry conviction

You must somehow convince your reader that the product is what you claim it to be. You must support your claims by evidence – facts, opinions. You can do this in a number of ways, for example:

- invite the reader to your factory or showroom
- offer to send goods on approval
- provide a guarantee
- quote your years of experience in the field

In this extract from a letter from a cotton shirt manufacturer, note how convincing it sounds. No manufacturer would dare to make such an offer without the firm belief in what is claimed. Your letter must give readers complete confidence and persuade them to buy

> Remember, we have manufactured cotton shirts for 50 years and are quite confident that you will be more than satisfied with their quality.
>
> This offer is made on the clear understanding that if the goods are not completely to your satisfaction you can return them to us without any obligation whatever and at our own expense. The full amount you paid will be refunded immediately.

As a caution, however, it is against the law to make false or exaggerated claims. Remember also that the good name and standing of your business, as well as its success, depends upon honest dealing.

Encourage action

Your closing paragraph must persuade the reader to take the action that you want – to visit your showrooms, to receive your representative, to send for a sample or to place an order.

The final paragraph must also provide readers with a sound reason why they should reply

> If you return the enclosed request card we will show you how you can have all the advantages of cold storage and at the same time save money.

Sometimes the closing paragraph will give special reasons why the reader should act immediately

> The special discount now offered can be allowed only on orders placed by 30 June. So hurry and take advantage of this limited offer while there is still time.

You must make it easy for the reader to do these things, such as providing a tear-off slip to complete and return or enclosing a prepaid card.

When composing sales letters, remember: your readers will not be anywhere near as interested in the product, service or idea you have to sell as they are in how it will benefit them. You must persuade your readers of the benefits of what you are selling and tell them what it can do for them.

 TIP Avoid 'hype' (hyperbole), eg totally fantastic, truly awesome, extraordinary, incredible, astounding.

Specimen sales letters

Here are some examples of effective sales letters that follow the four-point plan of *interest*, *desire*, *conviction* and *action*.

Sales appeal to economy

Dear Mr Wilson

Attention —— Have you ever thought how much time your typist wastes in taking down your dictation? It can be as much as a third of the time spent on correspondence. If you record your dictation – on our *Dictagram* – she can be doing other jobs while you dictate.

Interest —— You will be surprised at how little it costs. For 52 weeks in the year, your *Dictagram* works hard for you, and you can never give it too much to do – all for less than an average month's salary for a secretary! It will take dictation anywhere at any time – during lunch-hour, in the evening, at home – you can even dictate while you are travelling or away on business. Simply mail the recorded messages back to your secretary for typing.

Desire —— The *Dictagram* is efficient, reliable, time-saving and economical. Backed by our international reputation for reliability, it is in regular use in thousands of offices all over the country. It gives superb reproduction quality, with every syllable as clear as a bell. It is unbelievably simple to use – just slip in a tiny preloaded cassette, press a button, and it is ready to record your dictation, interviews, telephone conversations, reports, instructions or whatever.

Conviction —— Nothing could be simpler! And with our unique after-sales service contract you are assured lasting operation at the peak of efficiency.

Action —— Some of your business friends are sure to be using our *Dictagram*. Ask them about it before you place an order and we are sure they will back up our claims. If you prefer, return the enclosed prepaid card and we will arrange for our representative to call and arrange a demonstration for you. Just state the day and time that will be most convenient for you.

Yours sincerely

Sales appeal to efficiency

Dear Mr Wood

Attention —— RELIANCE – EVERY CAR OWNER'S DREAM!

Interest —— Reports from all over the world confirm what we have always known – the RELIANCE solid tyre is the fulfilment of every car owner's dream.

Desire —— You will naturally be well aware of the weaknesses of the ordinary air-filled tyre – punctures, outer covers which split under sudden stress, and a tendency to skid on wet road surfaces, to mention only a few of motorists' main complaints. Our RELIANCE tyre enables you to offer your customers a tyre which is beyond criticism in the essential qualities of road-holding and reliability.

Conviction —— We could tell you a lot more about RELIANCE tyres but would prefer you to read the enclosed reports from racing car drivers, test drivers, motor dealers and manufacturers. These reports speak for themselves.

Action —— To encourage you to hold a stock of the new solid RELIANCE, we are pleased to offer you a special discount of 3% on any order received by 31 July.

Yours sincerely

Sales appeal to security

Dear Mr Goodwin

A client of mine is happier today than he has been for a long time – and with good reason. For the first time since he married 10 years ago, he says he feels really comfortable about the future. Should he die within the next 20 years, his wife and family will now be provided for. For less than £3 a week paid now, his wife would receive £100 per month for a full 20 years, and then a lump sum of £20,000.

Such protection would have been beyond his reach a short time ago, but a new scheme has enabled him to ensure this security for his family. The scheme does not have to be for 20 years. It can be for 15 or 10 or any other number of years. And it need not be for £20,000. It could be for much more or much less so that you arrange the protection you want.

For just a few pounds each month you can buy peace of mind for your wife, your children and for yourself. You cannot – you dare not – leave them unprotected.

I would welcome an opportunity to call on you to tell you more about this scheme, which so many families are finding so attractive. I shall not press you to join; I shall just give you all the details and leave the rest to you. Please return the enclosed prepaid reply card and I will call at any time convenient to you. Alternatively, you could visit our website **www.greaterlife.co.uk**, click on contact, and complete the simple form.

I look forward to hearing from you.

Yours sincerely

Sales appeal to comfort

Dear Mrs Walker

What would you say to a gift that gave you a warmer and more comfortable home, free from draughts, and a saving of over 20% in fuel costs?

You can enjoy these advantages, not just this year but every year, simply by installing our SEALTITE panel system of double glazing. Can you think of a better gift for your entire family? The enclosed brochure will outline some of the benefits that make

SEALTITE the most completely satisfactory double-glazing system on the market thanks to a number of features not provided in any other system.

The panels are precision-made by experienced craftsmen to fit your own particular windows. Remember, you will be dealing with a well-established company that owes its success to the satisfaction given to thousands of customers.

There is no need for you to make up your mind right now. Why not let us give you a free demonstration in your own home without any obligation? If you are looking for an investment with an annual average return of over 20%, here is your opportunity. If you post the enclosed card to reach us by the end of August, we can complete the installation for you in good time before winter sets in. Or, even quicker, just click on **www.yorksealtite.co.uk** and click on the special banner at the top, type in the code 20SEAL, and someone will be in touch with you soon.

Secure your home with SEALTITE!

Yours sincerely

Sales appeal to leisure

Dear Mrs Hudson

'Modern scientific invention is a curse to the human race and will one day destroy it', said one of my customers recently. Rather a hasty statement, and quite untrue. There are modern inventions which, far from being a curse, are real blessings.

Our new AQUAMASTER washer is just one of them. It takes all the hard work out of the weekly wash and makes washing a pleasure. All you have to do is put your soiled clothes in, press a button and sit back while the machine does the work. It does every-thing – washing, rinsing and drying – and we feel it does it quicker and better than any washing machine on the market today.

Come along and see the AQUAMASTER at work in our showroom. A demonstration will take up only a few minutes of your time, but it may rid you of your dread of wash-ing day and make life much more pleasant.

I hope you will accept this invitation and come along soon to see what this latest of domestic time-savers can do for you.

Yours sincerely

TIP Stress the benefits of the product or service rather than the features.

Sales appeal to sympathy

This letter was sent together with a leaflet containing a form for readers to return with a donation.

Dear Reader

You can walk about the house, at work, in the streets, in the country. You take this ability for granted, yet it is denied to thousands of others – those who are born crippled, or crippled in childhood by accident or illness.

It is estimated that every five minutes in Britain a deformed child is born or a child is crippled by accident or illness. This means that every day there could be 288 more crippled children.

Does this not strike you as unfair? Most of what is unfair in life is something we can do little about but here is one very important inequality which everyone can help with. The enclosed leaflet explains how you can help. Please read it carefully while remembering again just how lucky you are.

Yours faithfully

Another sales appeal to comfort

Dear Homeowner

Great news! At half the actual cost, you can now have Solar Heating installed in your home.

As part of our research and development scheme introduced two years ago, we are about to make our selection of a number of properties throughout the country as 'Research Homes' – and yours could be one of them.

The information received from selected 'Research Homes' in the past two years has proved that Solar Heating is successful even in the most northern parts of the United Kingdom. This information has also enabled us to modify and improve our designs, which we will continue to do.

If your home is selected as one of the properties to be included in our research scheme, we will bear half the actual cost of installation.

If you are interested in helping our research programme in return for a half-price Solar Heating system, please complete the enclosed form and return it by the end of May. Within three weeks we will inform you if your home has been selected for the scheme.

Yours sincerely

Sales appeal to heat

Dear Madam

Thousands of people who normally suffer from the miseries of cold, damp, changeable weather wear THERMOTEX. Why? The answer is simple – tests conducted at the leading Textile Industries Department at Leeds University have shown that of all the traditional underwear fabrics THERMOTEX has the highest warmth insulating properties.

THERMOTEX has been relieving aches and pains for many years, particularly those caused by rheumatism. It not only brings extra warmth but also soothes those aches caused by icy winds cutting into your bones and chilling you to the marrow. THERMOTEX absorbs much less moisture than conventional underwear fabrics, so perspiration passes straight through the material. It leaves your skin dry but very, very warm.

Don't just take our word for it – take a good look at some of the testimonials shown in the enclosed catalogue. The demand for THERMOTEX garments has grown so much in recent years that we often have to deal with over 20,000 garments in a single day.

The enclosed catalogue is packed with lots of ways in which THERMOTEX can keep you warm and healthy this winter. Just browse through it, choose the garment you would like, and send us your completed order form – our FREEPOST address means there is even no need for a postage stamp!

Warmth and health will soon be on their way to you. If you are not completely satisfied with your purchase, return it to us within 14 days and we will refund your money without question and with the least possible delay.

Let THERMOTEX keep you warm this winter!

Yours faithfully

CAUTION In your sales letters, don't talk badly about your competition. Let your product sell itself.

Voluntary offers

An offer that is not asked for and which is sent to an individual or a small number of individuals is a form of sales letter. It serves the same purpose and follows the same general principles.

Such offers take a variety of forms. They may be:

- offers of free samples
- goods on approval
- special discounts on orders received within a stated period
- offers to send brochures, catalogues, price lists, patterns, etc, upon return of a form or card which is usually prepaid.

Offer to a newly established trader

Dear Sir

We would like to send our best wishes for the success of your new specialised toyshop. I know you will wish to offer your customers the latest toys – toys that are attractive, hard wearing and reasonably priced. Your stock will not be complete without the mechanical toys for which we have a national reputation.

We are sole importers of *Valifact* toys and our terms are very generous, as you will see from the enclosed price list. In addition to the trade discount stated, we will allow you a special first-order discount of 5%.

We hope these terms will encourage you to place an order with us, and feel sure you would be well satisfied with your first transaction.

We will be happy to arrange for one of our representatives to call on you to ensure that you are fully briefed on the wide assortment of toys we can offer. Please complete and return the enclosed card to say when it would be convenient.

We are looking forward to working with you.

Yours faithfully

Offer to a regular customer

Dear Mr Welling

We have just bought a large quantity of high quality rugs and carpets from the bankrupt stock of one of our competitors.

As you are one of our most regular and long-standing customers, we would like you to share in the excellent opportunities our purchase provides. We can offer you mohair rugs in a variety of colours at prices ranging from £55 to £3,000, also premier quality Wilton and Axminster carpeting in a wide range of patterns at 20% below current wholesale prices.

This is a fantastic opportunity for you to buy a stock of high-quality products at prices we cannot repeat. We hope you will take full advantage of it.

If you are interested, please call our hotline number 0104 2348282 not later than next Friday 14 October and we'll be happy to show you the stock for yourself. Or alternatively, call our Sales Department on 0114 4532567 to place an immediate order.

Yours sincerely

Offer to new homeowners

Dear Newcomers

Welcome to your new home! We have no wish to disturb you as you settle in but we would like to tell you why people in this town and the surrounding areas are very familiar with the name *Baxendale*.

Baxendale is situated at the corner of Grafton Street and Dorset Road, and we invite you to visit us to see for yourself the exciting range of goods that have made us a household name.

Our well-known shopping guide is enclosed for you to browse through at your leisure. You will see practically everything you need to add to the comfort and beauty of your home.

As a special attraction to newcomers to the area, we are offering a free gift worth £5 for every £50 spent in our store. The enclosed card is valid for one calendar month and it will entitle you to select goods of your own choice as your free gift.

We hope that you enjoy living in your new home, and look forward to seeing you soon.

Yours faithfully

Introduction to new product

Aurora Mobile
it's important to stay . . . in touch

Dear Valued Customer

With the introduction of our new **Connect Card**, Aurora Mobile has brought a new era of convenience in mobile communications. With the **Connect Card** you can enjoy all the benefits of Aurora Mobile's leading-edge network without worrying about monthly bills. Find out how in this month's issue of **In Touch**.

In Touch also introduces you to our vastly expanded international roaming services – Roam-a-round. This allows you to roam to all corners of the globe. Inside **In Touch** you will find out why no one covers the world better than Aurora Mobile.

Many more features can be found inside **In Touch**:

- Generous savings when you call another Aurora Mobile customer.
- What's new at our website.
- See and read about our performance at a recent Communications Exhibition.

Inside **In Touch** we have also included an exciting contest for you to win fabulous prizes. We have a free subscription to our value-added services, a free **Connect Card** worth £20 and restaurant privileges in leading restaurants.

With your continued support we have become the UK's leading network service provider. Thank you for staying with us.

Yours sincerely

Sabrina Chan (Ms)
Senior Director
Marketing, Sales and Customer Service

Aurora Mobile, Aurora House, Temple Street, London SE1 4LL
Tel: +44(0)181 542 4444 Fax: +44(0)181 555 4444
Email: auroramobile@cfb.co.uk

Offer of a demonstration

Dear Mrs Thornton

The Ideal Home Exhibition opens at Earls Court on Monday 25 June, and you are certain to find attractive new designs in furniture as well as many new ideas.

The exhibition has much to offer that I know you will find useful. We would like to invite you to visit our own display on Stand 26 where we shall be revealing our new *Windsor* range of unit furniture.

Windsor represents an entirely new concept in luxury unit furniture at very modest prices and we hope you will not miss the opportunity to see it for yourself. The inbuilt charm of this range comes from the use of solid elm and beech, combined with expert craftsmanship to give a perfect finish to each piece of furniture.

I enclose two admission tickets to the Ideal Home Exhibition. I am sure you will not want to miss this opportunity to see the variety of ways in which *Windsor* unit furniture can be arranged to suit any requirements.

I look forward to seeing you there.

Yours sincerely

 TIP Errors lurk in the middle of long sentences. Keep them short!

Useful expressions

Openings

- We are enclosing a copy of our latest catalogue and price list.
- As you have placed many orders with us in the past, we would like to extend our special offer to you.
- We are able to offer you very favourable prices on some goods we have recently been able to purchase.
- We are pleased to introduce our new ___ and feel sure that you will find it very interesting.
- I am sorry to note that we have not received an order from you for over ___.

Action/closes

- We hope you will take full advantage of this exceptional offer.
- We feel sure you will find a ready sale for this excellent material and that your customers will be well satisfied with it.
- We would be pleased to provide a demonstration if you would let us know when this would be convenient.
- We feel sure you will agree that this product is not only of the highest quality but also very reasonably priced.

 Checklist

- Arouse interest in first paragraph.
- Create a desire for the product or service.
- Describe the product if appropriate.
- Point out the benefits.
- Stress quality and special features.
- Convince the reader that what you claim is correct.
- Give evidence to support your claim.
- Persuade the reader to take appropriate action.

Publicity material

26

How to write a punchy press release

A press release is a statement prepared for distribution to the media. Its main purpose is to give journalists information that is useful, accurate and interesting. Every organisation has news that will interest potential customers, such as:

- relocation of offices
- expansion of business
- a new product launch
- a current promotion
- a local team sponsorship
- changes in top personnel
- human interest stories.

You can send your press (or media) release to local or national newspapers, or to other media like radio and TV stations. There are also many services available online now that will distribute your press release to every major news site and search engine on the Internet, and can put you in front of consumers and journalists.

If you visit the website of any organisation, you will most probably find a tab 'News releases' or 'Media'. There you will find all their news releases, with hyperlinks taking readers to connecting stories or sites of interest.

For your release to be considered newsworthy it must have a legitimate news angle (announcing something new and/or timely). It must not contain obvious sales language, and it must have a broad general interest or a special angle. It must be written objectively and in third person, as though someone else is writing the story.

 CAUTION Remember that you are giving information, not just selling something. You must aim to show the editor or reporter the newsworthiness of a particular event, service, product or person.

Formatting your press release

Journalists receive a lot of press releases every day, so they have expectations that you must conform to if you want your release to be read, let alone published.

While no one can guarantee that your press release will be published or used for an article, there are things you can do to improve your chances. The biggest obstacle to most press releases is the release itself. Here are some guidelines:

1 Main headings

Press releases should be printed on company letterhead in double line spacing. At the top spell out the words PRESS RELEASE in capitals and bold. If the press release is for IMMEDIATE RELEASE, say so, on the left margin directly above the title in capitals. Otherwise, you may state the date when the information may be released. This is called an EMBARGO DATE.

2 Compose a compelling headline

Just as the subject line of an email should attract the reader to open a message, a press release headline should pack such a punch that the reader wants to read on. It should be brief, clear and straight to the point – a really compact version of the story's key point. Remember, this headline may be your only chance. Don't try to write the headline until you've finished the body of the release though.

3 Draw the reader in with the 'lead'

The first sentence of the body is called the lead. This must hook the reader in and make him want to keep reading.

4 Structure the body of the release logically

■ After the lead, the next paragraph must actually sum up the press release. Journalists won't read the entire press release if the start of the article doesn't generate interest.

■ Keep the body copy compact. Use simple sentences and short paragraphs. Don't try to use fancy language and always avoid jargon.

- Provide as many actual facts as possible – events, products, services, people, targets, goals, plans, projects.
- Communicate the 5 Ws and the H: who, what, when, where, why and how. Address these issues in your news story:

  What is the actual news?

  Why is this news?

  The people, products, items, dates and other things related with the news

  The purpose behind the news

  Your company – the source of this news

- The more newsworthy you make the press release copy, the better the chance it will be selected by a journalist for reporting.
- Always include a quotation – ideally from the main person involved in the item. The text need not be an actual quote but it should be realistic. Always check that the person being quoted is happy with it. A quotation allows a busy journalist to prepare a complete article without doing a follow-up interview.

5 Include a powerful call to action

What action would you like people to take after reading? Make sure you invite them to take that action. Perhaps you want them to buy a product? If so, include information on where the product is available. Maybe you want them to enter a contest on your website, or learn more about your organisation? If so, include URL or a phone number.

6 Use anchor text links

Anchor text links are essential in press releases. Make sure all your keywords are hyperlinked to pages on your website or blog that are optimised for the same keywords. This is a contributing factor to overall search engine marketing success. This way, when press release syndication networks distribute your content, the keywords will still be hyperlinked to the content you're promoting. Remember too, people read content, so it's important to ensure that the keywords make sense in the structure and flow of your copy.

7 Include contact details

If your press release is really newsworthy, journalists will want more information and may like to interview key people. Always include:

The company's name

Contact person's name and title

Office address

Telephone and fax numbers

Mobile phone numbers

Email addresses

Websites

8 Signal the end

It's usual practice to signal the end of a press release with three # signals, centred directly underneath the last line.

 TIP It's a good idea to research actual press releases on the Internet to get a feel for the tone, the language, the structure and the format of press releases. Many organisations have a media or press section.

Top 10 tips for writing a press release

1 Make sure the information is newsworthy.

2 Start with a brief description of the news, then indicate who announced it, not the other way around.

3 Ask yourself, 'How are people going to relate to this and will they be able to connect?'

4 Make sure the first 10 words of your release are effective, as they are the most important.

5 Show your audience that the information is intended for them and make it compelling reading.

6 Avoid excessive use of adjectives and elaborate language.

7 State just the key facts without rambling.

8 Provide as much contact information as possible: person's name, address, phone, fax, email, website address.

9 Make sure you have a great issue to report on in your press release.

10 Word your document so well that it is easy for the editor or reporter to do their jobs.

 TIP A press release is not an advertisement. Be sure to write as though you are the editor of the newspaper speaking, in third person.

Press release announcing new store

① Turner Communications   Mobile Phone specialists
21 Ashton Drive
Sheffield   Tel +44 114 2871126
S26 2ES   Fax +44 114 2871123
                               Email TurnerComm@intl.uk

Include reference and date — ST/BT

15 June 201-

The embargo date is the date before which the information cannot be published — EMBARGO DATE: Immediate

NEW JOBS IN TURNER SUPERSTORE

Introduction: state the main message quickly — Mobile phone specialists, Turner Communications, have today announced the opening of their new store Turner's Office Supplies. More than 50 new jobs have been created.

Use short, self-contained paragraphs – include all essential details — Turner Communications have established themselves as leaders in the field of mobile communications in the UK. Roaming agreements have been set up with many countries throughout the world.

Use double spacing for the press release — The company has now announced that it is diversifying. Their new Office Supplies superstore will sell everything from stationery and office sundries to computers and other office equipment. It will be situated in a prime location at Meadowhall Retail Park on the outskirts of Sheffield, very close to the M1 motorway.

A grand opening ceremony is planned for Monday 1 July with special offers for the first 100 customers and a grand draw at 5.00 pm.

Round it off with a conclusion or quotation — Sally Turner, Managing Director, said, 'We are very excited about this new office superstore and feel confident that it will prove to be an overwhelming success.'

State contact details (for further information/ photographs) — Contact: Susan Gingell, Marketing Manager,
             Turner Communications
             Telephone: 0114 2871126

Press release announcing new hotel wing

FK/ST

14 September 201-

EMBARGO DATE: Immediate

NEW SERVICE CONCEPTS AT PAGODA SINGAPORE

Service, the magic word in today's hotel industry, gains a new perspective when the new Regency Suites wing of the Pagoda Hotel Singapore opens in early 201-.

The hotel's new upmarket product is targeted at the corporate traveller. In line with this, a range of personalised services in major areas can now be expected by the discerning traveller.

The Business Centre, a vital facility for businesspeople on the move, will operate 24 hours, 7 days a week. With this extension of operating hours, busy executives will enjoy the convenience of conducting business at any time of the day. Whether it is an urgent fax required at 1 am or an email, fax or letter in the middle of the night, time is no longer an issue. The Business Centre is well equipped with a complete range of secretarial services, including a comprehensive reference library, personal computer, access to the Internet, private offices and conference room with lounge.

Housekeeping and laundry services will also be available 24 hours daily. Guests arriving late at night will no longer worry about getting a suit pressed for the next morning. Requests for extra pillows, shampoo or stationery, or any other item will be met regardless of the hour.

A professional concierge team will answer queries and provide the wealth of information often required, from dinner reservations to theatre shows, or even finding the best shoemaker in town.

The hotel's airport representatives will not only greet guests upon arrival at the airport but also meet them during departure too. In addition to its 2 limousines, a fleet of 14 other cars are available at all times for a city tour or business trips.

With 148 Pagoda Hotels and resorts around the world, the Pagoda Singapore is positioning itself as a top deluxe hotel, making it the perfect choice for any traveller.

###

Contact: Florence Cheung, Public Relations Manager
            Telephone 3432343 Extension 145

TIP A flat, dull, boring, long-winded press release will end up in the editor's waste-paper bin!

Newsletters

Many organisations publish regular newsletters to keep staff informed about matters of interest. They may be called staff magazines or in-house journals. These newsletters are a good way of keeping employees informed and improving company/staff relations. They are an effective way of reaching out to staff where there are many different branches of a company around the country or internationally. The newsletters may also be sent to retired staff.

Company newsletters may contain a variety of information, such as:

- new policies/procedures
- updates on products/services
- developments in certain industries
- news from branches/departments
- promotions
- births/marriages/deaths
- retirements
- sports and social news
- successes and achievements
- contributions from employees
- calendar items/dates to remember
- health and stress tips.

Some companies produce separate newsletters for customers or members of a special industry. They are a good way of keeping people informed about the latest products, news and developments within the company or industry.

Article in staff newsletter

Catchy headline —— SUPERSTARS TEAM GAIN SECOND PLACE

Give main message —— The stamina and strength of three Global employees were put
quickly in the first   to the test when they competed in the European finals of the
paragraph   Tech-stars competition held in Rotterdam, Holland.

Use short, self- —— Global Holdings was invited to the European finals after winning
contained paragraphs   the regional heat at Leeds and being runners-up in the British final.

Use double spacing in —— All entrants must work with information technology in some way,
case of editing   and Global has entered a team every year since 1985 when they
                        won the European final. This year's competition consisted of eight
                        strenuous, athletic-based events in one day, in which three of the
                        five team members had to compete.

Make it a human- —— Unfortunately, due to holiday commitments, this year's Global
interest story   team entered without two of their top athletes, leaving John
                        Holmes, Martin Wilson and Andrew Johnson to compete in this
                        event. After a long day's work the team then had to face the final
                        event, which was a 2000 metre steeplechase, and all team
                        members performed extremely well in this.

Finish with a snappy —— The final result was that Global put in a very creditable
conclusion   performance and achieved second place. Well done to the team!

Email newsletters

E-newsletters have become very common place. Organisations, associations and companies of all sizes and types produce e-newsletters. Unfortunately, these e-newsletters contribute to our 'inbox clutter'.

However, it must be said that e-newsletters can still be a powerful marketing tool. They enable companies to maintain a relationship with customers, as well as to foster relationships with people who aren't quite ready to buy. E-newsletters position your company as an industry expert, are far more cost-effective than printed mail, and are easy to create. But if they are to stand out from the rest of the 'inbox clutter', they will have to be well crafted.

The bottom line is that email newsletters can be a very effective marketing tool, as long as you use some creativity, and focus on your readers' needs.

Here are a few tips for creating e-newsletters that will attract and retain readers instead of having them search for 'unsubscribe':

1 Use a professional and appropriate subject line. 'ABC Company's Newsletter' is way too boring. Subject lines that include phrases such as 'How to . . .', '10 Tips for . . .', 'Secrets of . . .', '8 Successful . . .', 'Top Tips for . . .', 'Mistakes To Avoid When . . .' will grab your readers' attention instead of driving them to the 'delete' key.

2 Keep it short. Provide headlines and brief paragraphs that link to the full articles on your website or blog. Readers may then scan the content quickly, and follow the links to articles of interest.

3 Provide news that is interesting to your reader rather than lots of company news.

4 Include news about industry trends or statistics – buying patterns, products, regulations. The more industry-focused, the better.

5 Offer tips and advice. People love that.

6 Include stories about successful business people that will inspire your readers.

7 Make it interactive by including a simple topic relating to an industry (eg 'How effective are webinars for you?'). People love to do such polls and see how the results shape up.

8 Let your voice be heard. Express an opinion about a topic of interest, similar to what bloggers do.

9 Include some fun. Business is serious, so a lighthearted touch is often appreciated. Perhaps you could include a cartoon, a trivia question, a quote of the day, or something else amusing.

10 Maintain consistency in format, tone and delivery frequency.

Writing skills

The same writing skills are needed for articles for magazines or e-newsletters:

HEADLINE   Compose an interesting, snappy headline that tells the whole story in a line.

OPENING   Write a good opening paragraph to grab the editor's or reader's attention. Give the main gist of the message here.

MIDDLE
- Use short, self-contained paragraphs
- Write in the third person as if the editor is 'speaking'
- Use short sentences and a crisp, snappy style
- Be as factual as possible
- Build the article logically
- Remember the 5 Ws:
  - What is happening?
  - Who is involved?
  - Where is it happening?
  - When is it happening?
  - Why is it newsworthy?

CLOSING   A quotation from a key person is very useful to close, otherwise a summary or conclusion.

Article announcing dinner and dance

ARE YOU READY FOR THE GLOBAL DINNER AND DANCE?

The year has flown and it's time once again to get ready for the Global Annual Dinner and Dance. Put these details in your diary now:

Where?	Dynasty Suite, Shangri La Hotel
When?	Saturday 17 December 201-
What time?	7.30 pm until late

As usual there will be a 10-course Chinese dinner (we can of course cater for any special requirements). Carmen Fashions will be entertaining us with a fashion show as we eat. With lucky draws, spot prizes and after-dinner entertainment and dancing, it's sure to be a great evening that you will not want to miss.

This company-sponsored dinner dance will cost you only S$50 each. Partners pay the same price too. If it's anything like previous years' functions, you can be assured of a fabulous time. Get your registration forms from Reception or the Human Resource Department – and remember to book early.

If you have any queries please contact:

Caroline Marshall
Human Resource Department
Extension 216
Email: carolinemarshall@global.com.my

Article about safety

Punchy headline —— SAFETY FIRST

Give the main gist in the —— A two-day workshop on Health and Safety in Laboratories was
opening paragraph   held at the group and regional laboratory in York in March.

Short, self-contained —— Staff from factories across the country attended the workshop,
paragraphs   organised by ST Training Solutions in collaboration with Ian Burke
    from the group occupational health and safety department.

Give full details and —— The speakers considered the legal and company requirements
information   for safety matters, including risk assessments, accidents and their
    prevention, fire safety, dealing with spillages in laboratories, and
    the correct procedures for arranging for disposal of waste.

Humour helps —— In one of the practical exercises, Ian Burke 'volunteered' to use
    the emergency shower, to check that it worked properly and to
    demonstrate that it is quite difficult to remove clothing whilst
    standing under a torrent of cold water!

It's a good idea to finish —— Says Ian, 'The social side of the workshop was enlivened by
with a quotation   joining one of the ghost tours of York in the evening, when we
    heard stories of York's macabre past!'

ST/JKL
24 October 201-

TIP Always look critically at everything you write, and remember to
proofread carefully.

Checklist

- Compose a snappy headline for any article or press release, and limit this to no more than one line – it should really grab the editor's attention and provide an interesting snapshot of what the article is all about.

- Grasp the reader's attention with a good opening paragraph, including all the key details. Get to the point quickly.

- Keep central paragraphs short and self-contained, so that the editor may cut them out if necessary. Make sure the details flow in a logical sequence.

- Make sure you include all the details like who, what, when, where, why and how.

- Use an interesting, snappy, punchy style for your writing. Even a seemingly uninteresting event can be made into an effective, appealing story by clever wording.

- Write in an objective style, as if the writer has no affiliation with your company, as if the newspaper or magazine is actually 'speaking'.

- Close by saying something exciting about the main message again. A quotation from a key person is very useful in this final section.

- Remember to include appropriate contact details: at least a name, website URL and telephone number.

Business plans

If you are setting up a new business or want people to invest serious money in your business, you need a business plan. A business plan defines your business and identifies your operations and your aims. It provides specific and organised information about your company. Basically, a business plan enables a business to look ahead, allocate resources, focus on key points, and prepare for challenges and opportunities.

A business plan has two main purposes:

1. To provide you with a detailed plan to help you as you make your new venture grow.
2. To convince investors that you are the sort of person and that this is the sort of business in which they should invest.

Why do you need a business plan?

Your business plan will be useful in many ways:

1. It will help you define your business and focus on your aims.
2. It can be used as a selling tool when you have to approach lenders, investors and banks.
3. Going through the process of compiling your business plan will help you uncover weaknesses or omissions in your planning processes.
4. You can use the plan to solicit opinions and advice from people, including those in your intended field of business, who will be able to give you advice.

What to avoid in your business plan

1 It may be a good idea to limit your long-term or future projections, because very often after a year or two the reality of your business may be quite different to your initial concept. Short-term objectives will be more realistic. You can then modify your plan as your business develops.

2 Keep your language simple, and explanations brief.

3 Remember to spell out strategies in case of business adversities. Some people ignore this.

Is there a standard template for a business plan?

Business plan formats vary, but generally a plan includes various components such as a description of the company, product or service, the market, forecasts and predictions, details of the management team, and some financial analysis.

Your own business plan will depend on your specific situation. Make sure your plan matches its purpose. Of course there are many helpful resources on the Internet where you can view real sample business plans and get some good ideas on how to compile your own plan.

Components of a business plan

Before writing your business plan it will help to look at as many examples as possible. Sample plans are available for all types of companies. Most of them suggest that your business plan should include similar components.

1 Executive summary

This is the first section of your business plan, and it is exactly what it sounds like – a compact, concise outline of the whole plan. In this section you should state:

- the nature of the company
- the products/services you will offer
- what's special about your products/services
- who the managers are
- how much money you need and what you will use it for.

Most people only read this executive summary, so you need to generate excitement here, make it sound interesting, and show how unique your business and your team are.

2 Table of contents

Try to keep this to one page, listing everything your plan includes along with page numbers.

3 Company description

- How did you get started?
- How has the company grown?
- Provide a history of sales, profits and other important information.
- Where are you now?
- What plans do you have for the future?

4 Products/services

Put yourself in the investor's shoes and ask yourself what you would want to know before investing money in your business. Focus on customer benefits, and make sure it answers questions like these:

- What products or services do you offer?
- What makes them different from others?
- How does it make people's lives better?
- What kind of equipment do you need?

5 Market analysis

This section shows all the research you have done, such as distribution problems, government regulations, technological opportunities, industry characteristics and trends, projected growth, customer behaviour, complementary products/services, etc.

6 Marketing plan

Start with a discussion of what the market is like. Where are your customers? What are their needs? How will you reach your customers and capture the market? List the steps you will take to ensure that customers know about your product/service and why they will prefer it over the competition. List all the tactics you will use, from the cheapest to the most expensive.

7 Operations plan (strategy and implementation)

This is the nuts and bolts of your business plan. You need to give precise information about what is involved in running your business. You may break it down into sections including your value proposition, marketing strategy, positioning statements, pricing strategy, promotion strategy, distribution patterns, marketing programmes, sales strategy, sales forecasts, strategic alliances, etc.

8 Web plan summary

Here you must include details of your website marketing strategy and development requirements.

9 Financial plan

In this section you will need to include details of sales forecasts, profit-and-loss statements, cash flow projections, balance sheets, etc.

 CAUTION Make sure you have accurate and up-to-date figures in your business plan. This will be essential if you are to obtain support.

10 Management

Include details of the management of your company, who is on the board, who will manage each department and why.

11 Appendices

There will probably be lots of appendices attached to your business plan – managers' résumés, promotional materials, product photographs and descriptions, financial details, etc.

Checklist

- Read lots of sample business plans – just put in a search on any search engine and you will find plenty of useful sites.

- Check out the bookshelves too – there are lots of books providing excellent detailed advice on how to write your business plan.

- Don't wait until the last minute to start writing. If you have ideas for your business, a solid plan will enable you to formulate all your thoughts better.

- Concentrate on strengths rather than originality. Investors usually pay more attention to the strength of the management team than to looking for a truly original idea.

- Try to be as concise as possible. Business plans are, of essence, very long documents, probably 40–60 pages.

- Keep focused. Stick to the essential facts, and cut out any padding.

- Spruce up your display. Make an effort to jazz up the formatting with headings, shaded sections, tables, charts, bullets and other graphics.

- Package it appropriately. It's not worth spending a lot of money on expensive leather-bound packaging. Investors want something that is easy to read and that lies flat on the desk.

- Edit. Edit. Edit. Get it right and you will achieve the results you want.

- Proofread the executive summary carefully. This is the most important section – make sure it is exciting, interesting and that there are no mistakes. People should read this and want to know more!

Appendix 1

Spoken and written forms of address

This section provides the correct forms of address for many officials, diplomats, religious leaders, royalty and the British peerage. The chart gives the appropriate form or forms to be used in addressing letters, in salutations, in direct conversation and in more formal introductions.

In diplomatic and other public circles, 'Sir' is generally considered an acceptable alternative to the formal address in both written and spoken greetings; this does not apply to religious or titled persons. The use of 'Madam' or 'Ma'am' for a female is less customary but still acceptable, especially for high officeholders ('Madam Governor'). This rule also holds for high officials of foreign countries.

Person	Letter address	Letter greeting	Spoken greeting	Formal introduction
President of the United States	The President The White House Washington, DC 20500	Dear Mr (or Madam) President	Mr (or Madam) President	The President or the President of the United States
Former President	The Honorable Jack Kimball Address	Dear Mr Kimball	Mr Kimball	The Honorable Jack Kimball
Vice President	The Vice President Executive Office Building Washington, DC 20501	Dear Mr (or Madam) Vice President	Mr (or Madam) Vice President	The Vice President or the Vice President of the United States
Cabinet members	The Honorable John (or Jane) Smith The Secretary of XXX or The Attorney General Washington, DC	Dear Mr (or Madam) Secretary	Mr (or Madam) Secretary	The Secretary of XXX

Person	Letter address	Letter greeting	Spoken greeting	Formal introduction
Chief Justice	The Chief Justice The Supreme Court Washington, DC 20543	Dear Mr (or Madam) Justice or Dear Mr (or Madam) Chief Justice	Mr (or Madam) Chief Justice	The Chief Justice
United States Senator	The Honorable John (or Jane) Smith United States Senate Washington, DC 20510	Dear Senator Smith	Senator Smith	Senator Smith from Nebraska
Ambassador	The Honorable John (or Jane) Smith Ambassador of the United States American Embassy Address	Dear Mr (or Madam) Ambassador	Mr (or Madam) Ambassador	The American Ambassador The Ambassador of the United States of America
Consul-General	The Honorable John (or Jane) Smith American Consul-General Address	Dear Mr (or Mrs, Ms) Smith	Mr (or Mrs, Ms) Smith	Mr (or Mrs, Ms) Smith
Foreign Ambassador	His (or Her) Excellency John (or Jane) Smith The Ambassador of XXX Address	Excellency or Dear Mr (or Madam) Ambassador	Excellency; or Mr (or Madam) Ambassador	The Ambassador of XXX
Secretary-General of the United Nations	His (or Her) Excellency Jack (or Jane) Smith Secretary-General of the United Nations United Nations Plaza New York, NY 10017	Dear Mr (or Madam) Secretary-General	Mr (or Madam) Secretary-General	The Secretary-General of the United Nations

Person	Letter address	Letter greeting	Spoken greeting	Formal introduction
Governor	The Honorable Jack (or Jane) Smith Governor of XXX State Capitol Address	Dear Governor Smith	Governor or Governor Smith	The Governor of Maine: Governor Smith of Washington
State legislators	The Honorable Jack (or Jane) Smith Address here	Dear Mr (or Mrs, Ms) Smith	Mr (or Mrs, Ms) Smith	Mr (or Mrs, Ms) Smith
Judges	The Honorable John Smith Justice, Appellate Division Supreme Court of the State of XXX Address	Dear Judge Smith	Justice or Judge Smith; Madam Justice or Judge Smith	The Honorable Jack (or Jane) Smith; Mr Justice Smith or Judge Smith; Madam Justice Smith or Judge Smith
Mayor	The Honorable Jack (or Jane) Smith; His (or Her) Honor the Mayor City Hall Address	Dear Mayor Smith	Mayor Smith; Mr (or Madam) Mayor; Your Honor	Mayor Smith; The Mayor
The Pope	His Holiness, the Pope or His Holiness, Pope John XII Vatican City Rome, Italy	Your Holiness or Most Holy Father	Your Holiness or Most Holy Father	His Holiness, the Holy Father; the Pope; the Pontiff
Cardinals	His Eminence, Martin Cardinal Brown, Archbishop of XXX Address	Your Eminence or Dear Cardinal Brown	Your Eminence or Cardinal Brown	His Eminence, Cardinal Brown
Bishops	The Most Reverend Martin Brown, Bishop (or Archbishop) of XXX Address	Your Excellency or Dear Bishop (Archbishop) Brown	Your Excellency or Bishop (Archbishop) Brown	

Person	Letter address	Letter greeting	Spoken greeting	Formal introduction
Monsignor	The Reverend Monsignor Nigel Frangoulis Address	Reverend Monsignor or Dear Monsignor	Monsignor Frangoulis or Monsignor	Monsignor Frangoulis
Priest	The Reverend Jack Smith Address	Reverend Father or Dear Father Smith	Father or Father Smith	Father Smith
Brother	Brother Jack or Brother Jack Smith Address	Dear Brother Jack or Dear Brother	Brother Jack or Brother	Brother Jack
Sister	Sister Linda Wright	Dear Sister Linda Wright or Dear Sister	Sister Linda Wright or Sister	Sister Linda Wright
Protestant clergy	The Reverend John (or Jane) James	Dear Dr (or Mr, Ms) James	Dr (or Mr, Ms) James	The Reverend (or Dr) Jack James
Bishop (Episcopal)	The Right Reverend Jack James Bishop of XXX Address	Dear Bishop James	Bishop James	The Right Reverend Jack James, Bishop of XXX
Rabbi	Rabbi Arnold (or Amanda) Schwartz Address	Dear Rabbi Schwartz	Rabbi Schwartz or Rabbi	Rabbi Arnold Schwartz
King or Queen	His (Her) Majesty King (Queen) XXX Address (letters traditionally are normally sent via the private secretary)	Your Majesty; Sir or Madam	Varies depending on titles, holdings, etc	
Other royalty	His (Her) Royal Highness, the Prince (Princess) of XXX	Your Royal Highness	Your Royal Highness; Sir or Madam	His (Her) Royal Highness, the Duke (Duchess) of XXX

Person	Letter address	Letter greeting	Spoken greeting	Formal introduction
Duke/ Duchess	His/Her Grace, the Duke (Duchess) of XXX	My Lord Duke/ Madam or Dear Duke of XXX/ Dear Duchess	Your Grace or Duke/Duchess	His/Her Grace, the Duke/ Duchess of XXX
Marquess/ Marchioness	The Most Honourable the Marquess (Marchioness) of Newport	My Lord/Madam or Dear Lord/ Lady Newport	Lord/Lady Newport	Lord/Lady Newport
Earl	The Right Honourable the Earl of Bangor	My Lord or Dear Lord Bangor	Lord Bangor	Lord Bangor
Countess (wife of an Earl)	The Right Honourable the Countess of Bangor	Madam or Dear Lady Bangor	Lady Bangor	Lady Bangor
Viscount/ Viscountess	The Right Honourable the Viscount (Viscountess) Manson	My Lord/Lady or Dear Lord/ Lady Manson	Lord/Lady Manson	Lord/Lady Manson
Baron/ Baroness	The Right Honourable Lord/ Lady Grey	My Lord/Madam or Dear Lord/ Lady Grey	Lord/Lady Grey	Lord/Lady Grey
Baronet	Sir Jack Smith, Bt	Dear Sir or Dear Sir Jack	Sir Jack	Sir Jack Smith
Wife of Baronet	Lady Smith	Dear Madam or Dear Lady Smith	Lady Smith	Lady Smith
Knight	Sir Elton John	Dear Sir or Dear Sir Elton	Sir Elton	Sir Elton John
Wife of Knight	Dear Madam or Dear Lady John	Lady John	Lady John	

Appendix 2
The A–Z of alternative words

Calling all writers

This guide gives hundreds of plain English alternatives to the pompous words and phrases that litter official writing. On its own the guide won't teach you how to write in plain English. There's more to it than just replacing 'hard' words with 'easy' words, and many of these alternatives won't work in every situation. But it will help if you want to get rid of words like 'notwithstanding', 'expeditiously' and phrases like 'in the majority of instances' and 'at this moment in time'. And using everyday words is an important first step towards clearer writing.

Copyright

Plain English Campaign owns the copyright on this guide. You must not copy it without getting our permission first. You can download your own copy from our website (**www.plainenglish.co.uk**).

Using the A to Z

If you find yourself about to write, type or dictate a word you wouldn't use in every day conversation, look it up in the A to Z. You should find a simpler alternative. Often there will be a choice of several words. You need to pick the one that best fits what you are trying to say.

A

(an) absence of	no, none
abundance	enough, plenty, a lot (or say how many)
accede to	allow, agree to
accelerate	speed up
accentuate	stress
accommodation	where you live, home

accompanying	with
accomplish	do, finish
according to our records	our records show
accordingly	in line with this, so
acknowledge	thank you for
acquaint yourself with	find out about, read
acquiesce	agree
acquire	buy, get
additional	extra, more
adjacent	next to
adjustment	change, alteration
admissible	allowed, acceptable
advantageous	useful, helpful
advise	tell, say (unless you are giving advice)
affix	add, write, fasten, stick on, fix to
afford an opportunity	let, allow
afforded	given
aforesaid	this, earlier in this document
aggregate	total
aligned	lined up, in line
alleviate	ease, reduce
allocate	divide, share, add, give
along the lines of	like, as in
alternative	choice, other
alternatively	or, on the other hand
ameliorate	improve, help
amendment	change
anticipate	expect
apparent	clear, plain, obvious, seeming
applicant (the)	you
application	use
appreciable	large, great
apprise	inform, tell
appropriate	proper, right, suitable
appropriate to	suitable for
approximately	about, roughly
as a consequence of	because
as of the date of	from
as regards	about, on the subject of
ascertain	find out

assemble	build, gather, put together
assistance	help
at an early date	soon (or say when)
at its discretion	can, may (or edit out)
at the moment	now (or edit out)
at the present time	now (or edit out)
attempt	try
attend	come to, go to, be at
attributable to	due to, because of
authorise	allow, let
authority	right, power, may (as in 'have the authority to')
axiomatic	obvious, goes without saying

B

belated	late
beneficial	helpful, useful
bestow	give, award
breach	break
by means of	by

C

calculate	work out, decide
cease	finish, stop, end
circumvent	get round, avoid, skirt, circle
clarification	explanation, help
combine	mix
combined	together
commence	start, begin
communicate	talk, write, telephone (be specific)
competent	able, can
compile	make, collect
complete	fill in, finish
completion	end
comply with	keep to, meet
component	part
comprise	make up, include
(it is) compulsory	(you) must
conceal	hide
concerning	about, on

conclusion	end
concur	agree
condition	rule
consequently	so
considerable	great, important
constitute	make up, form
construe	interpret
consult	talk to, meet, ask
consumption	amount used
contemplate	think about
contrary to	against, despite
correct	put right
correspond	write
costs the sum of	costs
counter	against
courteous	polite
cumulative	added up, added together
currently	now (or edit out)
customary	usual, normal

D

deduct	take off, take away
deem to be	treat as
defer	put off, delay
deficiency	lack of
delete	cross out
demonstrate	show, prove
denote	show
depict	show
designate	point out, show, name
desire	wish, want
despatch or dispatch	send, post
despite the fact that	though, although
determine	decide, work out, set, end
detrimental	harmful, damaging
difficulties	problems
diminish	lessen, reduce
disburse	pay, pay out
discharge	carry out
disclose	tell, show

disconnect	cut off, unplug
discontinue	stop, end
discrete	separate
discuss	talk about
disseminate	spread
documentation	papers, documents
domiciled	in living in
dominant	main
due to the fact of	because, as
duration	time, life
during which time	while
dwelling	home

E

economical	cheap, good value
eligible	allowed, qualified
elucidate	explain, make clear
emphasise	stress
empower	allow, let
enable	allow
enclosed	inside, with
(please find) enclosed	I enclose
encounter	meet
endeavour	try
enquire	ask
enquiry	question
ensure	make sure
entitlement	right
envisage	expect, imagine
equivalent	equal, the same
erroneous	wrong
establish	show, find out, set up
evaluate	test, check
evince	show, prove
ex officio	because of his or her position
exceptionally	only when, in this case
excessive	too many, too much
exclude	leave out
excluding	apart from, except
exclusively	only

exempt from	free from
expedite	hurry, speed up
expeditiously	as soon as possible, quickly
expenditure	spending
expire	run out
extant	current, in force
extremity	limit

F

fabricate	make, make up
facilitate	help, make possible
factor	reason
failure to	if you do not
finalise	end, finish
following	after
for the duration of	during, while
for the purpose of	to, for
for the reason that	because
formulate	plan, devise
forthwith	now, at once
forward	send
frequently	often
furnish	give
further to	after, following
furthermore	then, also, and

G

generate	produce, give, make
give consideration to	consider, think about
grant	give

H

henceforth	from now on, from today
hereby	now, by this (or edit out)
herein	here (or edit out)
hereinafter	after this (or edit out)
hereof	of this
hereto	to this

heretofore	until now, previously
hereunder	below
herewith	with this (or edit out)
hitherto	until now
hold in abeyance	wait, postpone
hope and trust	hope, trust (but not both)

I

if and when	if, when (but not both)
illustrate	show, explain
immediately	at once, now
implement	carry out, do
imply	suggest, hint at
in a number of cases	some (or say how many)
in accordance	with as under, in line with, because of
in addition (to)	and, as well as, also
in advance	before
in case of	if
in conjunction with	and, with
in connection with	for, about
in consequence	because, as a result
in excess of	more than
in lieu of	instead of
in order that	so that
in receipt of	get, have, receive
in relation to	about
in respect of	about, for
in the absence of	without
in the course of	while, during
in the event of/that	if
in the majority of instances	most, mostly
in the near future	soon
in the neighbourhood of	about, around
in view of the fact that	as, because
inappropriate	wrong, unsuitable
inception	start, beginning
incorporating	which includes
incurred	have to pay, owe

indicate	show, suggest
inform	tell
initially	at first
initiate	begin, start
insert	put in
instances	cases
intend to	will
intimate	say, hint
irrespective of	despite, even if
is in accordance with	agrees with, follows
is of the opinion	thinks
issue	give, send
it is known that	I/we know that

J

jeopardise	risk, threaten

L

(a) large number of	many, most (or say how many)
(to) liaise with	to meet with, to discuss with, to work with (whichever is more descriptive)
locality	place, area
locate	find, put

M

magnitude	size
(it is) mandatory	(you) must
manner	way
manufacture	make
marginal	small, slight
material	relevant
materialise	happen, occur
may in the future	may, might, could
merchandise	goods
mislay	lose
modification	change
moreover	and, also, as well

N

negligible	very small
nevertheless	but, however, even so
notify	tell, let us/you know
notwithstanding	even if, despite, still, yet
numerous	many (or say how many)

O

objective	aim, goal
(it is) obligatory	(you) must
obtain	get, receive
occasioned by	caused by, because of
on behalf of	for
on numerous occasions	often
on receipt of	when we/you get
on request	if you ask
on the grounds that	because
on the occasion that	when, if
operate	work, run
optimum	best, ideal
option	choice
ordinarily	normally, usually
otherwise	or
outstanding	unpaid
owing to	because of

P

(a) percentage of	some (or say what percentage)
partially	partly
participate	join in, take part
particulars	details, facts
per annum	a year
perform	do
permissible	allowed
permit	let, allow
personnel	people, staff
persons	people, anyone
peruse	read, read carefully, look at

place	put
possess	have, own
possessions	belongings
practically	almost, nearly
predominant	main
prescribe	set, fix
preserve	keep, protect
previous	earlier, before, last
principal	main
prior to	before
proceed	go ahead
procure	get, obtain, arrange
profusion of	plenty, too many (or say how many)
(to) progress something	(replace with a more precise phrase saying what you are doing)
prohibit	ban, stop
projected	estimated
prolonged	long
promptly	quickly, at once
promulgate	advertise, announce
proportion	part
provide	give
provided that	if, as long as
provisions	rules, terms
proximity	closeness, nearness
purchase	buy
pursuant to	under, because of, in line with

Q

qualify for	can get, be able to get

R

reconsider	think again about, look again at
reduce	cut
reduction	cut
referred to as	called
refer to	talk about, mention
(have) regard to	take into account
regarding	about, on

regulation	rule
reimburse	repay, pay back
reiterate	repeat, restate
relating	to about
remain	stay
remainder	the rest, what is left
remittance	payment
remuneration	pay, wages, salary
render	make, give, send
represent	show, stand for, be
request	ask, question
require	need, want, force
requirements	needs, rules
reside	live
residence	home, where you live
restriction	limit
retain	keep
review	look at (again)
revised	new, changed

S

said/such/same	the, this, that
scrutinise	read (look at) carefully
select	choose
settle	pay
similarly	also, in the same way
solely	only
specified	given, written, set
state	say, tell us, write down
statutory	legal, by law
subject to	depending on, under, keeping to
submit	send, give
subsequent to/upon	after
subsequently	later
substantial	large, great, a lot of
substantially	more or less
sufficient	enough
supplement	go with, add to
supplementary	extra, more
supply	give, sell, delivery

T

(the) tenant	you
terminate	stop, end
that being the case	if so
the question as to whether	whether
thereafter	then, afterwards
thereby	by that, because of that
therein	in that, there
thereof	of that
thereto	to that
thus	so, therefore
to date	so far, up to now
to the extent that	if, when
transfer	change, move
transmit	send

U

ultimately	in the end, finally
unavailability	lack of
undernoted	the following
undersigned	I, we
undertake	agree, promise, do
uniform	same, similar
unilateral	one-sided, one-way
unoccupied	empty
until such time	until
utilisation	use
utilise	use

V

variation	change
virtually	almost (or edit out)
visualise	see, predict

W

ways and means	ways
we have pleasure in	we are glad to

whatsoever	whatever, what, any
whensoever	when
whereas	but
whether or not	whether
with a view to	to, so that
with effect from	from
with reference to	about
with regard to	about, for
with respect to	about, for
with the minimum of delay	quickly (or say when)

Y

you are requested	please
your attention is drawn	please see, please note

Z

zone	area, region

Words and phrases to avoid

The words and phrases below often crop up in letters and reports. Often you can remove them from a sentence without changing the meaning or the tone. In other words, they add nothing to the message. Try leaving them out of your writing. You'll find your sentences survive and succeed without them.

- a total of
- absolutely
- abundantly
- actually
- all things being equal
- as a matter of fact
- as far as I am concerned
- at the end of the day
- at this moment in time
- basically

- current
- currently
- during the period from
- each and every one
- existing
- extremely
- I am of the opinion that
- I would like to say
- I would like to take this opportunity to
- in due course
- in the end
- in the final analysis
- in this connection
- in total
- in view of the fact that
- it should be understood
- last but not least
- obviously
- of course
- other things being equal
- quite
- really
- really quite
- regarding the (noun), it was
- the fact of the matter is
- the month(s) of
- to all intents and purposes
- to one's own mind
- very

Appendix 3
How to write reports in plain English

Introduction

Welcome to the plain English report-writing course. All you need is a pen, some paper, a little time and the will to learn.

There is no great mystery about writing clear, concise and effective reports. The writing skills you will learn in this book will work in all types of 'business' writing – letters, leaflets, memos and so on. What makes reports different is the formal way they are organised, and we'll be looking at that.

All the examples are genuine.

The answers to all the exercises are at the end of the guide.

At the end of the course is a list of common bureaucratic words with plain English alternatives.

Copyright

Plain English Campaign owns the copyright on this guide. You must not copy it without getting our permission first. You can download your own copy from our website (**www.plainenglish.co.uk**).

So what's plain English?

First let's say what plain English isn't and destroy some of the myths about it.

- It's not 'cat sat on the mat' or 'Peter and Jane' writing. Almost anything – from leaflets and letters to legal documents – can be written in plain English without being patronising or over-simple.

- It doesn't mean reducing the length or changing the meaning of your message. Most of the UK's biggest insurance companies produce policies that explain everything fully in plain English.

- It's not about banning new words, killing off long words or promoting completely perfect grammar. Nor is it about letting grammar slip.
- It is not an amateur's method of communication. Most forward-looking senior managers always write in plain English.
- And finally, it is not as easy as we would like to think.

Sadly, thanks to the bureaucrats of public service industries, local councils, banks, building societies, insurance companies and government departments, we have learned to accept an official style of writing that is inefficient and often unfriendly.

But in the last few years, many of these offenders have started to put things right, either rewriting their documents clearly or training their staff in the art of plain English or both.

The advantages of plain English are:

- it is faster to write;
- it is faster to read; and
- you get your message across more often, more easily and in a friendlier way.

If you spend more than an hour a day writing, you are (to an extent) a professional writer. So it's vital that you get it right.

Plain English Campaign has led the way in the field of clear communication. The Campaign edits and designs documents for the country's largest organisations and runs hundreds of training courses every year. Now Plain English Campaign has used all their experience to put together this teach-yourself course on writing reports in plain English.

So what is plain English? It is a message, written with the reader in mind and with the right tone of voice, that is clear and concise.

Keep your sentences short

We're not going to join in the argument about 'what is a sentence?'. Just think of it as a complete statement that can stand by itself. Most experts agree that clear writing should have an average sentence length of 15–20 words.

This does not mean making every sentence the same length. Be punchy. Vary your writing by mixing short sentences (like the last one) with longer ones (like this one), following the basic principle of sticking to one main idea in a sentence, plus perhaps one other related point. You should soon be able to keep to the average sentence length – used by top journalists and authors – quite easily.

At first you may still find yourself writing the odd long sentence, especially when trying to explain a complicated point. But most long sentences can be broken up in some way.

Exercise

Here are some examples. Split them where suitable by putting in full stops. You may need to put in or take out words so that the new sentences will make sense. But don't change anything else.

1 From a formal report of a disciplinary interview

I raised your difficulty about arriving ready for work on time and pointed out that your managers had done their best to take account of your travel problems and you had agreed with them that the Green Lane depot was the most convenient place for you to work, however, your initial improvement was short-lived and over the past two months your punctuality has dropped to a totally unacceptable level.

2 From an electricity company

I do not seem to have received the information required from you to set up your budget scheme, and I now enclose the relevant form and ask that you fill it in and return it.

3 From a solicitor

If you could let me have the latest typed version of the form in the next seven days, whereupon I suggest we meet here on 19 December to finalise the text so that you could then give me an estimate of the cost of producing a typeset proof.

4 From a credit company

I refer to the earlier notice served in respect of your account as the arrears now amount to the sum shown above, you leave me with no alternative than to commence court action and details of your account have been referred to the company's solicitor.

Actives and passives

Do you want your reports to sound active or passive – crisp and professional or stuffy and bureaucratic?

Well, this is where we have to get grammatical. Most people know that a verb is a 'doing' word, like 'make', 'do', 'play', 'talk' or 'write'. There are many ways to split verbs into different categories, but we're just going to consider the difference between active and passive verbs.

Passive verbs make writing duller and more difficult to understand. Active verbs make writing livelier and more personal.

But what are active and passive verbs? Let's take a simple sentence: 'The boss slammed the door.'

Here, we can call the boss 'the doer'. The verb is 'slammed'. And the door is what we can call 'the thing'.

In almost all sentences that contain active verbs, the doer comes first, then the verb and then the thing. There will probably be lots of other words as well. For example: 'The boss, in a fit of temper, slammed the door to the outer office.' But the order of doer, verb, thing stays the same.

With passive verbs, the thing comes first: 'The door was slammed by the boss.' You can see that by making the sentence passive, we have had to introduce the words 'was' and 'by', and the sentence becomes more clumsy.

Remember that the doer is not always a person and the thing is not always a thing! 'The tree crushed Peter' is active but 'Peter was crushed by the tree' is passive. And remember 'passive' has nothing to do with the past tense.

Here are some more examples of sentences containing passive verbs. Our 'active' versions are underneath each one.

- **The matter will be considered by us shortly.**
  (We will consider the matter shortly.)
- **The riot was stopped by the police.**
  (The police stopped the riot.)
- **The mine had to be closed by the safety inspector**.
  (The safety inspector had to close the mine.)

Sometimes the doer gets left out.

Sentences with passive verbs can make sense without having a doer. For instance, 'the door was slammed', 'the cheque had been cashed' and 'the report is being written' all leave out the doer.

People used to officialese often write reports that are full of passive verbs, with sentences like these.

■ 35 sites were visited in three weeks. Procedures were being properly followed at the sites visited.

■ Overheads were not kept under control despite an awareness of the budgetary situation.

Neither of these sentences has a doer. So the reader may be left asking, 'Who visited the sites?', 'Who was following procedures properly?' and so on. Changing to active verbs reveals the 'doers' and sharpens up dull and unclear sentences.

■ We visited 35 sites in three weeks. At the sites we visited, we found that staff were following procedures properly.

■ Managers were not keeping overheads under control, despite knowing about the budgetary situation.

You will notice that in the last sentence we have used an active verb instead of 'an awareness of'. As we shall see later this is an example of changing a 'nominalisation' into a verb.

Spotting passives

There is another way of spotting passive verbs which is especially useful when the doer isn't mentioned in the sentence.

First, passive verbs almost always have one of the following words added on – be, being, am, are, is, was, were, will be. They are all formed from the verb 'to be'.

Second, they have a thing called a 'past participle'.

This table shows you how to get a past participle from a verb.

Verb	Past participle
ask	asked
claim	claimed
do	done
write	written

So a complete passive verb could be 'will be done', 'has been formed' or 'was watched'.

Here are some examples:

■ Care should be taken when opening the door.
■ The outcome will be decided next week.

- Applicants will only be accepted if proof of purchase is enclosed.
- It can be done. The problem could not have been foreseen.

Good uses of passives

There are times of course when it makes sense to use a passive.

- To make something less hostile – 'this bill has not been paid' (passive) is softer than 'you have not paid this bill' (active).
- To avoid taking the blame – 'a mistake was made' (passive) rather than 'we made a mistake' (active).
- When you don't know who or what the doer is – 'the England team has been picked'.
- If it simply sounds better.

But aim to make about 80–90% of your verbs active.

Exercise

The difference between active and passive verbs is not easy to grasp. So if you are confused, read this section again. If you are not, spot the passive verbs in the following examples and change the sentences around so that they use active verbs.

1 From a DVLA letter (you will need to invent a doer for the first verb)

The tax disc was sent to you at the address on your application form but it was returned by the Post Office as undeliverable mail.

2 From a building society

In the Investment Account Statement which was sent to you recently, it was indicated by us that we would write to you again concerning the monthly interest that has been paid to you under the terms of your account.

3 From a council leaflet to parents (use 'we' for the Education Department, 'you' for the parent)

Advice must also be sought from any other professional likely to have relevant information. If there is anyone whom you think should be consulted, for example a specialist doctor your child is seeing, please let the Area Education Office know. Every professional whose advice is sought will be sent a copy of any information that is provided by you.

Talk to your reader

Write with your reader in mind. If you want to encourage people to read your report, give them a piece of writing that is lively and readable. Imagine you are presenting your report to your reader yourself. Think carefully: What do they know already? What do you need to tell them? Talk directly to your readers in language they understand. You will find that using shorter sentences and active verbs will already have made a difference.

'I' and 'we'

As we said earlier, reports used to be full of passive verbs. This allowed the writer to remain anonymous by leaving out the doer. They used phrases like:

- it was found that;
- it is accepted that; and
- it is recommended that.

The reason (or excuse) for this used to be that the writer was writing on behalf of the organisation. But usually everyone knows who has written the report, who interviewed people and so on.

Let the readers know there is a person behind the print. It's not just friendlier; 'I', 'you' and 'we' are also usually easier to understand. Use phrases like:

- we found that;
- I accept that; and
- we recommend that.

Obviously you will use 'I' if the report is all your own work, and 'we' if you are reporting on a team effort.

But be sure that your audience knows who you are talking about.

Understandable words

Say exactly what you mean, using the simplest words that fit. This does not necessarily mean only using simple words – just words that the reader will understand.

At the end of the course is a list of a few of the words that we suggest you avoid. But for most words you will have to decide yourself whether they are suitable. Most importantly, don't use jargon that is part of your working life unless you are writing to someone who uses the same jargon. If a teacher is writing to an education officer, the jargon word 'SATs' could be very useful in saving time and space. But when writing a report that parents will read, you wouldn't use it without explaining what it means.

In general, use everyday English whenever you can. Again, imagine you are presenting your report in person. Write to communicate, not to impress.

What is a nominalisation?

A 'nominalisation' is a type of 'abstract noun'. In other words, it is the name of something that isn't a physical object but a process, technique or emotion.

Nominalisations are formed from verbs.

For example:

Verb	Nominalisation
complete	completion
introduce	introduction
provide	provision
fail	failure
arrange	arrangement
investigate	investigation
use	utilisation

So what's wrong with them?

The problem is that writers often use nominalisations when they should use the verbs they come from. Like passive verbs, too many nominalisations make writing very dull and heavy-going.

Here are some examples of nominalisations, with our plain English versions underneath.

- **We had a discussion about the matter.**
  (We discussed the matter.)
- **The report made reference to staff shortages.**
  (The report referred to staff shortages.)
- **The decision was taken by the Board.**
  (The Board decided.)
- **The implementation of the policy has been done by a team.**
  (A team has implemented the policy.)

Exercise

Bring the following sentences to life by revealing the verbs hidden by nominalisations and making any changes you think are necessary.

1 From a letter on Housing Benefit

If you would like consideration to be given to your application, please send me your last five wage slips.

2 From an electricity company

Your meter is operated by the utilisation of tokens.

3 From a surveyor's report

We can solve the problem by the removal of the plaster to a height of one metre, the insertion of a new damp proof course and the introduction of suitable floor joists.

4 From a gas region

We have made an examination of your account and can tell you that application for budget payments at this late stage is still possible.

5 From a council to a building contractor

To cause minimum disruption to teaching provision, it is imperative that the school roof renewal is carried out by your company concurrently with your completion of the Special Needs Unit.

Cutting out useless words

Now you're going to put all the techniques you've learnt so far into practice, along with one other technique – cutting out useless words.

So with all these examples, shorten sentences, put in active verbs, use everyday English, make them more personal and direct, replace nominalisations and cut out useless words.

1 From an electricity board letter

The standing charge is payable in respect of each and every quarter.

2 From a bank

You will be sent a letter regarding current interest rates not less often than once a year.

3 From a credit company

Notice must be given of your intention to cancel the agreement a period of 30 days prior to your cancellation.

4 From a building society

Should you be unable to agree to the contents of the statement or you have any questions thereon, please write to this department at the address overleaf, enclosing your passbook or certificate and the statement.

5 From a management consultant

You are required to notify us immediately in the event of your unavoidable absence from work for sickness or any other reason and the attached note explains your obligations in this respect.

6 From a solicitor

We would advise that attached herewith is the entry form which has been duly completed and would further advise that we should be grateful if you would give consideration to the various different documents to which we have made reference.

7 From another electricity company

In consequence of the non-payment of the above-mentioned account, an employee will call at your home for the purpose of obtaining a meter reading and disconnecting the supply on 10 March.

8 From a local authority

If you are experiencing difficulty in meeting your rent payments and are not currently in receipt of Housing Benefit, you may qualify for help towards your rent under the Housing Benefit Scheme and details of this can be forwarded upon request. Alternatively, if you require advice regarding either your rent arrears or possible entitlement to Housing Benefit, please do not hesitate to contact me.

Other points to consider

Sounding positive

Always try to emphasise the positive side of things.

For example:

- If you don't send your payment, we won't be able to renew your membership of the scheme. (negative)

■ Please send your payment so that we can renew your membership of the scheme. (positive)

Now rewrite these examples in a positive way (and use what you have learned so far to make any other changes you think are necessary).

1 From a local authority

You will reduce your chances of a council home if you do not keep your choice of areas as wide as possible.

2 From a bank (make this more personal)

No-one may apply for the scheme unless their account is credited with at least £1,000.

Using lists

Lists are excellent for splitting information up.

There are two main types of list.

■ You can have a continuous sentence with several listed points picked out at the beginning, middle or end.

■ You can have a list of separate points with an introductory statement (like this list).

In the list above, each of the points is a complete sentence so they each start with a capital letter and end with a full stop.

For the same type of list with short points, it is better to set it out like this.

Kevin needed to take:

■ a penknife
■ some string
■ a pad of paper
■ a pen.

With a list that is part of a continuous sentence, put semi-colons (;) after each point and start each with a lower case letter.

If you can prove that:

■ you were somewhere else at the time;
■ you were not related to Mary; and
■ you are over 21;

then you should be all right.

As you can see, the next to last point has 'and' after the semi-colon. If you only had to prove one of the three points instead of all of them, this word would be 'or'.

Always make sure each point follows logically from the introduction. For example, if you took out 'you' from the second and third points it would still flow as a normal sentence but not as a list. The third point would effectively read 'If you can prove that are over 21' which obviously does not make sense.

We have also used bullet points for each listed point. These are better than numbers or letters as they draw your attention to each point without giving you extra information to take in.

Exercise

Use bullet points to split the following sentence into a list.

1 From a scientist's report (make this more personal)

People with a mathematical brain or those who display a very sharp sense of humour or chess players of club standard or higher are more likely to show early signs of being musical.

Some more myths destroyed

We're not trying to be trendy here by breaking some of the grammatical rules. We're just going to destroy some of the grammatical myths.

In the past, grammarians decided it was 'bad form' to do certain things. But our rule is, if it makes your sentence clearer, don't worry about 'bad form'.

You can start a sentence with 'and', 'but', 'because', 'so', 'or' or 'however'.

You can 'split infinitives'. So, you can say 'to boldly go'.

You can end a sentence with a preposition. In fact, it is something we should stand up for.

And you can repeat a word in a sentence if you can't find a better word.

Of course, this does not mean you should break these so-called 'rules' all the time – just when you want to make a sentence flow better.

Planning and organising reports

When writing reports, make your audience's job as easy as possible. Use active verbs, short sentences and keep to the point, just as you would in any other kind of writing.

First you need to plan and organise the report carefully. Before you write a report you need to:

- define its purpose carefully;
- investigate the topic thoroughly; and

Defining the purpose

This helps you to be clear about:

- why you are writing;
- what to include;
- what to leave out; and
- who your readers are.

If you can express the purpose in a single sentence, so much the better.

Investigating the topic

How you do this depends on the topic and purpose. You may need to read, interview, experiment and observe. Get advice from someone more experienced if you need to.

Organising the report into sections

Your job is to make it easy for the readers to find the information they want.

In reports that are one or two pages long, readers should have no trouble finding their way around. With a 'long' report (more than four or five pages), you need to take great care in how you organise the information.

Reports can be set out in eight parts, but you won't always need them all.

- Title or title page
- Contents list
- Abstract
- Introduction
- Discussion

■ Summary and conclusions

■ Recommendations

■ Appendix

A short report won't need a **title page**, but should have a title.

The **contents list** is only needed in long reports.

The **abstract** is only needed in formal reports, such as reports of scientific research. It is a summary of the report. The abstract appears in library files and journals of abstracts. It won't usually be printed with the report so it needs to be able to stand alone.

Keep it between 80 and 120 words. Don't confuse this with an 'executive summary' which we will talk about later.

The **introduction** should be brief and answer any of the following questions that seem relevant.

■ What is the topic?

■ Who asked for the report and why?

■ What is the background?

■ What was your method of working? If the method is long and detailed, put it in an appendix.

■ What were the sources? If there are many, put them in an appendix.

The **discussion** is the main body of the report. It is likely to be the longest section, containing all the details of the work organised under headings and sub-headings.

Few readers will read every word of this section. So start with the most important, follow it with the next most important, and so on.

You should follow the same rule with each paragraph. Begin with the main points of the paragraph, then write further details or an explanation.

The **summary and conclusions** section is sometimes placed before the discussion section. It describes the purpose of the report, your conclusions and how you reached them.

The conclusions are your main findings. Keep them brief. They should say what options or actions you consider to be best and what can be learned from what has happened before. So they may include or may lead to your recommendations: what should be done in the future to improve the situation?

Often, writers will put the summary and conclusions and the recommendations together and circulate them as a separate document. This is often called an executive summary because people can get the information they need without having to read the whole report.

It may be better (and cheaper) to send everyone an executive summary, and only provide a copy of the full report if someone asks for it. You may save a few trees, and you will certainly save your organisation plenty of time and money.

The **appendix** is for material which readers only need to know if they are studying the report in depth. Relevant charts and tables should go in the discussion where readers can use them. Only put them in an appendix if they would disrupt the flow of the report.

Order of presentation

We recommend the following order of presentation. You won't always need all these sections, especially those in brackets.

Long reports

- Title or title page
- (Contents list)
- (Abstract)
- Introduction
- Summary and conclusions
- Recommendations
- Discussion
- (Appendix)

Short reports

- Title
- Introduction
- Discussion
- Summary and conclusions
- Recommendations
- (Appendix)

The order in which you write needn't follow the order of presentation.

We recommend the following order of writing, because each section you finish helps you write the next one.

Order of writing

- Introduction
- Discussion
- Summary and conclusion
- Recommendations
- (Abstract)
- Title or title page
- (Contents list)
- (Appendix)

After writing all the sections, read and revise them. Rewrite sections if necessary.

Numbering sections and paragraphs

If you use plenty of clear headings and have a full contents list at the start of the report, you should find this is enough to show where each part begins and ends, and to cross-refer if necessary.

If you do have to label sections and paragraphs, keep it as simple as possible. Use capital letters to label sections and numbers to label paragraphs (A1, A2 and so on). If necessary, use small letters on their own for parts of paragraphs.

Planning the writing

Usually you will have collected such a mass of information that you cannot decide where to plunge in and begin. So, before you start to write you must make some kind of plan.

This will save you hours of writing and will help to produce a better-organised report.

Here are two different ways of planning.

An **outline** begins as a large, blank sheet of paper onto which you pour out all your facts, ideas, observations and so on, completely at random. Write in note form, and try to get everything down as fast as possible.

When you have got all your points on paper, start to organise them, group them, and assess them for strength, relevance, and their place in the report.

You can then number the points in order or put headings next to them such as 'Intro', 'Discussion', 'Conclusion' and so on. Use lines and arrows to link up related points.

Gradually you will create a network of ideas grouped under headings – this is the structure of your report. Leave it for a day or two if you can. Return with fresh ideas, add points you'd forgotten, and cross out anything you don't need.

Mind mapping is a different way of planning that suits some writing better. The idea is the same: by pouring out ideas at random, you can concentrate on the content, and organise the material at leisure when the ideas are set down.

There is no special magic to a mind map. Start by putting the topic in a box in the middle of the page, then draw lines to branch out from it with your main ideas.

It is easy to add new information and to make links between the main ideas. Order and organisation will often take care of themselves.

Revision

Always read critically what you've written. If possible, leave it alone for a few days and then re-read it. Or ask someone else to read it for you. Ask: 'Is this clear, concise and persuasive?' Be prepared to revise your language and structure. You may even have to rewrite parts that don't work.

Writing your report

This is your chance to practise everything you have learned, by planning and writing a report from scratch.

The purpose of the report

Your senior managers want to find out how staff travel to work and what influences their choice of travel. The results of the survey will help them make decisions about working hours, car parking and travel-to-work costs.

You have to prepare a short report on your own travel options and how and why you travel as you do. Write about 250 to 350 words (about a side of A4 or 12 to 18 sentences of 20 words).

If you live very close to work (or even work from home) you will have to use a little imagination. For instance, pretend you have moved several miles away, or that your employer has moved.

The method

Think about the purpose carefully, then produce a plan using a mind map or an outline. Do it quickly, pouring out the points as they occur to you. Don't worry about neatness – no-one else needs to be able to read your plan.

Take 10 to 15 minutes over this.

Look at your plan and assess the points for importance and relevance. Then organise them under section headings. You will probably need four sections.

- Title
- Introduction
- Discussion
- Summary and conclusions

Now write the report.

Take 20 to 25 minutes for this stage.

We haven't provided a 'model' answer here because everyone using this book will have different circumstances, and will write a different report.

But there are a number of ways you can check what you have written.

Have you used active verbs and the words 'I', 'me', 'my' and so on?

Because you were describing your own options and choices, this should have happened fairly naturally.

Is your average sentence length around 15 to 20 words?

You can check this by counting the number of words you have used, dividing that by the number of sentences.

In each section, and in each paragraph, have you given the most important information first, and then explained or given the detail?

Read the report out loud. Imagine you are telling your boss about your options for travel and so on. Are there any words or phrases in your written report that you wouldn't use if you were talking? What would you say instead? Could you use those words in your report?

'Brevity' – Memo to the War Cabinet

'To do our work, we all have to read a mass of papers. Nearly all of them are far too long. This wastes time, while energy has to be spent in looking for the essential points.

I ask my colleagues and their staff to see to it that their reports are shorter.

The aim should be reports which set out the main points in a series of short, crisp paragraphs.

If a report relies on detailed analysis of some complicated factors, or on statistics, these should be set out in an appendix.

Often the occasion is best met by submitting not a full-dress report, but an 'aide-memoire' consisting of headings only, which can be expanded orally if needed.

Let us have an end of such phrases as these:

'It is also of importance to bear in mind the following considerations', or 'Consideration should be given to the possibility of carrying into effect'. Most of these woolly phrases are mere padding, which can be left out altogether, or replaced by a single word. Let us not shrink from using the short expressive phrase, even if it is conversational.

Reports drawn up on the lines I propose may first seem rough as compared with the flat surface of officialese jargon. But the saving in time will be great, while the discipline of setting out the real points concisely will prove an aid to clearer thinking.'

Sir Winston Churchill, 9 August 1940

Summary

We hope you have enjoyed the course and that it will help you write in plain English.

- Plan carefully before you start writing. Use an outline or a mind map so that you know exactly what you'll be writing about.
- Organise your report into sections.
- Use everyday English whenever possible.
- Avoid jargon and legalistic words, and explain any technical terms you have to use.
- Keep your sentence length down to an average of 15 to 20 words. Try to stick to one main idea in a sentence.

- Use active verbs as much as possible. Say 'we will do it' rather than 'it will be done by us'.
- Be concise.
- Imagine you are talking to your reader. Write sincerely, personally, in a style that is suitable and with the right tone of voice.
- And always check that your report is accurate, clear, concise and readable.

Training from Plain English Campaign

We offer training courses to teach you how to design and write your documents in plain English. We run two types of course:

- open courses, held at various hotels throughout the country, where anyone can attend; and
- in-house courses, where we come to an organisation and train your staff. This means we can tailor our training to your organisation's work.

You can also follow our Plain English Diploma Course. This is a 12-month course, leading to a qualification in plain English.

We now offer two courses teaching English grammar. Our Grammarcheck Course is designed to teach delegates the fundamentals of grammar, punctuation, sentence construction and spelling which are so essential for clear communication. We also occasionally hold an Advanced Grammar Course, which goes into more detail on the grammar of standard English.

You may also be interested in The Plain English Course – our pack of materials to help you train your own staff.

For more details on any of the courses, please visit **www.plainenglish.co.uk/training.html.** If you have any specific questions about training courses, please call our training manager Terri Schabel on **01663 744409** or email us at **info@plainenglish.co.uk**

Words to avoid

Try to use the alternatives we suggest in brackets:

- additional (extra)
- advise (tell)

- applicant (you)
- commence (start)
- complete (fill in)
- comply with (keep to)
- consequently (so)
- ensure (make sure)
- forward (send)
- in accordance with (under, keeping to)
- in excess of (more than)
- in respect of (for)
- in the event of (if)
- on receipt (when we/you get)
- on request (if you ask)
- particulars (details)
- per annum (a year)
- persons (people)
- prior to (before)
- purchase (buy)
- regarding (about)
- should you wish (if you wish)
- terminate (end)
- whilst (while)

Our suggested answers

In some cases, our answers are about the only ones possible. However, most things can be expressed clearly and well in several different ways. If your answers are different from ours, don't worry. Compare them carefully and see whether there are any extra techniques you can pick up from them. And of course, our suggestions aren't perfect!

Keep your sentences short

1 I raised your difficulty about arriving ready for work on time. I pointed out that your managers had done their best to take account of your travel problems and you had agreed with them that the Green Lane depot was the most convenient place for you to work. However, your initial improvement was short-lived. Over the past two months your punctuality has dropped to a totally unacceptable level.

Or

I raised your difficulty about arriving ready for work on time. I pointed out that:

- your managers had done their best to take account of your travel problems; and

- you had agreed with them that the Green Lane depot was the most convenient place for you to work.

However, your initial improvement was short-lived. Over the past two months your punctuality has dropped to a totally unacceptable level.

2 I do not seem to have received the information required to set up your budget scheme. I now enclose the relevant form. Please fill it in and return it.

3 Please let me have the latest typed version of the form in the next seven days. I suggest we meet here on 19 December to finalise the text. You could then give me an estimate of the cost of producing a typeset proof.

4 I refer to the earlier notice served in respect of your account. As the arrears now amount to the sum shown above, you leave me no alternative but to commence court action. Details of your account have been referred to the company's solicitor.

Actives and passives

1 We sent you the tax disc at the address on your application form but the Post Office returned it as undeliverable mail.

2 In the Account Statement which we sent you recently, we indicated that we would write to you again concerning the monthly interest that we have paid to you under the terms of your account.

3 We must also seek advice from any other professional likely to have relevant information. If there is anyone you think we should consult, for example a specialist doctor your child is seeing, please let the Area Education Office know. We will send a copy of any information you provide to every professional whose advice we seek.

Talk to your reader

Please pay this debt immediately. You have four days to pay or to contact this office. (Ring the number shown above.) If you do not pay the debt or contact us within this time, we may have to start legal proceedings. So don't ignore this letter.

What is a nominalisation?

1. If you would like us to consider your application, please send me your last five wage slips.
2. You operate your meter using tokens.
3. We can solve the problem by removing the plaster to a height of one metre and putting in a new damp-proof course and suitable floor joists.
4. We have examined your account and can tell you that you can still apply for budget payments at this late stage.
5. To disrupt teaching as little as possible, your company must renew the school roof at the same time as finishing the Special Needs Unit.

Cutting out useless words

1. You must pay the standing charge every quarter.
2. We will send you a letter about current interest rates at least once a year.
3. You must tell us that you intend to cancel the agreement 30 days before you cancel.
4. If you don't agree with the contents of the statement or you have any questions about it, please write to us at the address over the page. Enclose your passbook or certificate and the statement.
5. You must tell us immediately if you are away from work for sickness or any other unavoidable reason. The attached note explains what you must do.
6. We have filled in the entry form and attached it to this letter. Please consider the various documents we have referred to.
7. As you have not paid this bill, we will call at your home on 10 March to cut off your supply and read your meter.
8. If you are having difficulty paying your rent and you do not get Housing Benefit, you may be able to get help towards your rent under the Housing Benefit Scheme. Please ask us for details. If you would like advice about your unpaid rent or whether you can get Housing Benefit, please contact me.

Other points to consider

1. You will increase your chances of a council home if you keep your choice of areas as wide as possible.
2. You may apply for the scheme if your account has £1,000 or more in it.
3. You are more likely to show early signs of being musical if you:
   - have a mathematical brain;
   - show a very sharp sense of humour; or
   - play chess to at least club standard.

Index